Building Services Procurement

Building Services Procurement

Christopher Marsh

LONDON AND NEW YORK

First published 2003
by Spon Press

2 Park Square, Milton Park, Abingdon, Oxfordshire OX14 4RN
711 Third Avenue, New York, NY 10017

Routledge is an imprint of the Taylor & Francis Group, an informa business

First issued in paperback 2018

Publisher's Note
This book has been produced from camera-ready copy supplied by the author

British Library Cataloguing in Publication Data
A catalogue record for this book is available from the British Library

Library of Congress Cataloging in Publication Data
A catalog record has been requested

ISBN 978-0-415-27477-7 (hbk)
ISBN 978-1-138-37249-8 (pbk)

For Maureen,
because she knows who she is.

The word craftsman can rarely be used with conviction. During the final stages of writing this book I learnt of the tragic death of one such individual. Ron Snow was one who shared with me his passion for design and workmanship. These lessons have remained with me during my career in construction.

This book is equally dedicated to the memory of Ron Snow.

Contents

Tables and Figures

Preface

This is a book about strategy. In the 1990s the subject of construction procurement matured and became a strategic issue. However, despite many publications, government reports and best intentions of the industry, the manner in which to best procure the services of a specialist engineering contractor and an engineering designer remains misunderstood. Given their value and increasing complexity, building services will nearly always dominate the project environment. Therefore, unless the industry moves to a greater level of understanding, the current drive for improved performance within the industry will go largely unfilled.

It would be arrogant to believe that this text would remedy this situation on its own. The text tries to outline the key issues that must be explored and understood while developing a suitable strategy. It is hoped that with the inclusion of modalities, a ready manner in which to implement the issues put forward will assist the reader.

To put the name of a single author on the front of a book is always selfish. It takes many people to realise the end goal of a completed text and these must be thanked: Tony Moore who first approached me with the idea and to Sarah Kramer and Alison Nick who provided support throughout the project: and for the many people who unselfishly provided information and assistance, including Gerry Samuelsson-Brown and Rohan Nanayakkara.

I am particularly grateful to Aileen Ryder who had the least enviable task of teaching an engineer grammar and turning my writings into a polished text. And finally to Erin, who on many Sunday afternoons reminded me that sometimes procurement has to take a back seat to playing on a Gameboy.

Christopher Marsh
April 2002

Chapter One

Introduction

1.1 THE ISSUE OF PROCUREMENT

During no other time in the history of the construction industry, has the subject of procurement dominated the debate on possible reforms. Regardless of country, from the United Kingdom, to Malaysia, Australia and America, the industry has been bombarded by material on how to improve the construction process. Procurement has often been the dominant issue, with the plethora of "new" procurement arrangements demonstrating the industry's and its clients' deep desire to seek improvements in the manner in which a project is formulated and executed.

Despite all of this, something is lacking. Very few of these industry reports or calls for improvements understand the complexity of initiating, then arranging, a supply chain as complex as that found on the average building services installation.

It would be arrogant to think that a single textbook will level this imbalance. Nevertheless, this text has been written to address the subject and cover as widely as possible the issues that have the greatest effect on the successful procurement of building services.

1.1.1 A Sub-optimality in Building Services Procurement

Procurement is defined as being to "Obtain by care or effort, acquire; bring about". In construction terms it is defined as "The framework within which construction is brought about, acquired or obtained". It is the term given to the process of organising and putting into action a new project. Determining a procurement strategy involves risk identification and assessing the most appropriate party to manage these risks. Although generic strategies exist, each can use contractual relationships to tailor a system specifically to meet a client's objectives.

This all sounds very simple and straightforward - needs are assessed, risks identified and apportioned, competent people are hired, a design is drawn and a competent contractor is hired to manage the works. So why does the procurement of a project seem to cause so many problems for building services?

This is mainly due to two highly complex reasons.

Firstly, each step is influenced by a large number of external factors, resulting in what academia calls a "wicked" problem. It is a problem with so many boundaries, external environments, subsystems, interactions and levels, that a single factor causes repercussions throughout the entire system. A simple example can be used to demonstrate this domino effect. If 20 subsystems are involved, and the first subsystem has one interaction with the next, which has two, and so on with the interactions doubling with each subsystem, by the time you reach the last subsystem you are dealing with 496,288 interactions: a truly wicked problem.

Secondly, traditionally procurement has been based on the tactics of traditional design-tender-construct mechanisms. Historically, building services contractors could be brought in at an early stage by the design team, on the basis that they would be nominated - a preliminary example of partnering. This simple process worked; however clients began to seek greater value for money and were adverse to taking risks, while consultants sought to discharge themselves from the risk of nominating a rogue contractor.

All of this has resulted in a lack of knowledge on how to procure building services in a strategic manner. Industry has ignored it due to nomination or a "lowest cost" wins mentality and academia has done so by only researching into nomination - an aversion to detailed research being driven by its wicked nature.

However, industry and academia have now woken up to the problem and some interesting developments have begun to emerge.

1.1.2 The Changing Nature of Project Procurement

Clients seeking a risk-free approach to their construction needs are largely driving the growing sophistication in construction procurement. The current trend in focused business has meant that most companies now outsource their non-core processes, in the quest for risk minimisation. This same philosophy has now spilt over into construction, with clients favouring procurement arrangements that either place all risk with the contractor or, through the use of investment vehicles, have the building provided to them as a serviced facility.

In the UK the government is leading changes in procurement regime. On 23 May 2000, the Treasury announced that in future all construction work, including maintenance and refurbishment, must be procured in one of three manners: design and build, prime contracting, or using the Private Finance Initiative. For building services, several key issues have now developed:

- the main contractor becomes the pseudo client
- the designer has a second tier seat
- choice can now be made on capital cost, whole life cost or operating cost: or any combination or permutation of these
- procurement of services has changed from the tactical to the strategic.

The supply chain must be joined together within a concept of a virtual network (in fact prime contracting demands this). Prime contracting requires that individual suppliers and contractors now deliver a complete product to the client. Modern strategies such as the UK's Ministry of Defence's prime contracting are based on this.

1.1.3 The Supplying Market

The marketplace that provides suitable organisations to participate within these modern arrangements is made up of a diversity of firms, but all of them are classified as small enterprises. Highlights of the UK market include:

- 137,000 companies supply a market worth £12 billion
- 99.5% of these companies employ less than 24 people
- the top 20 contractors control 27% of the market, with the largest firm controlling only 3.6%
- the median profit of all of these companies is 1.43%
- only 13% of tradesmen are directly employed.

The consultants market shows a similar picture:

- 2,729 professional services firms supply a design market of £720 million
- average profit is 6%
- 80% of firms are sole traders with a turnover of less than £99k
- 45% of smaller firms do not use information technology
- total employment of consultants equals 16,457
- firms are split over a range of 18 major disciplines - ranging from standard mechanical through to lifts, communications, fire engineering etc.

Therefore the supplying market is dominated by organisations that are small in nature, making minimal profits. This must be borne in mind when procuring building services. Due to the market, and diversity of technologies, even the simplest of projects will require multiple subcontracting.

This causes a de-facto approach to building services, for not only is there a diversity of organisations, but each specialist technology is often viewed as a separate trade or engineering profession

Mechanical	Electrical	Specialists
▪ air conditioning	▪ electrical distribution	▪ lifts and escalators
▪ heating / ventilation	▪ uninterrupted power	▪ kitchen equipment
▪ water services	▪ lighting	▪ commissioning
▪ smoke control	▪ lightening protection	▪ data cabling
▪ plumbing	▪ electric heating	▪ fire fighting
		▪ controls and BEMS
		▪ process/medial gas
		▪ security systems

Table 1.1 General Areas of Work Specialisation

This provides certain benchmarks for use when procuring building services:

- all projects will involve sub-contracting
- risk transfer to contractors is a false ideal as few have commercial reserves or buffers
- the design programme will be limited to the resources of the consultancy and the number of specialist designers
- all elements will require some form of design work, which will be split between manufactures, fabricators and the specialist contractor
- an introduction to systems theory.

We are now able to draw a composite picture of the problem that faces all procurers of building services. To be successful will require the balancing of many sub-systems comprising the various consultants, contractors and technical systems, all of which are operating in a diverse socio-organisational environment.

Building services have the following characteristics. They are:

- connected parts of other systems and environments;
- a complex whole of technical, organisational and institutional sub-systems;
- materially organised whole; and
- purposeful by having a change of inputs, synergy of parts and a definable boundary.

Furthermore, they are an open system as they respond to negativity, external environments and are continuously adapting to these influences.

Management is the process of pursuing effective and efficient activities with and through other people. It involves three simple processes:

- inquiry, through analysis that leads to planning
- action, through organising and leading
- cybernetics, through control and communication.

Therefore the procurement of building services can be seen as a pure management activity.

The ability to effectively plan and control the chaotic system that surrounds most management decisions led to the development of open systems theory in the 1950s. Success of a project, for example a new school, will be seen from a number of perspectives: from the students, from the faculty, the service manager, the contractors and so on. This is known as the world-view, and the various cultural divisions and objectives between these groups allows the different perspective. This now adds a third dimension to the problem: the system we are now dealing with becomes a web rather than linear.

The final dimension of the puzzle must centre on the critical success factors that are set for the project. At the end of the day, the client will be the only judge of the success of a project. The developed strategy must respect this and set clear measurable objectives in the form of critical success factors and defined processes to enable them through a clear strategy.

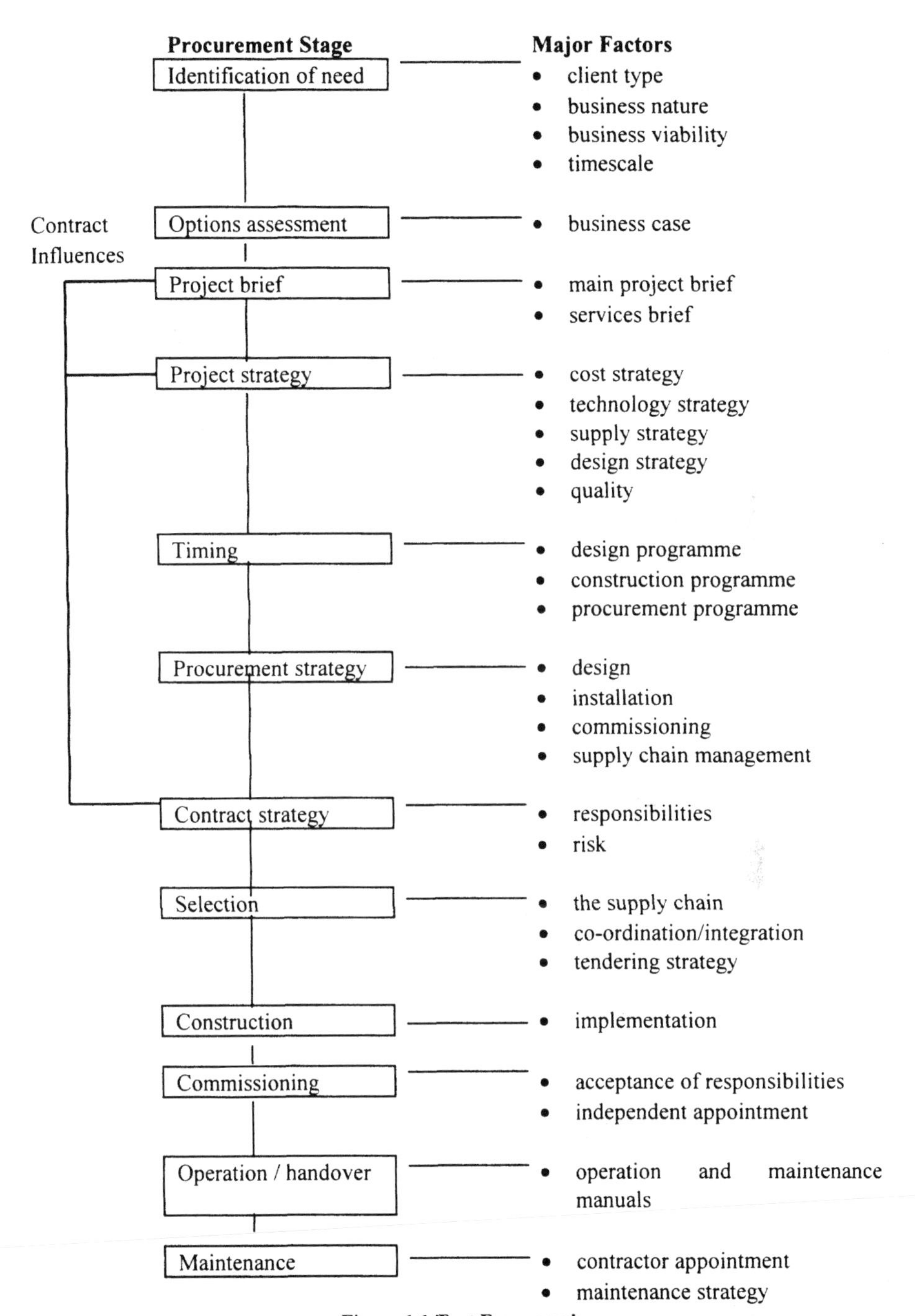

Figure 1.1 Text Framework

1.2 FRAMEWORK OF THE TEXT

The text approaches the subject from the science of operations management, where a number of key resources must be gathered and then with the aid of a series of management tools, each is utilised to its full extent. The delivery, being a co-ordinated approach to the development and final execution of a strategy, is focused on value delivery and maximisation of resource potential.

Regardless of the specific approach developed for the whole construction project, building services can be seen as a specific sub-project, where a separate strategy can be developed. Using the principles of supply chain management, a cornerstone in the operations management toolkit, a strategic approach can be developed that ensures the product remains focused on the clients needs, while the inherent issues of value and cost can be maximised and minimised respectively.

The expression "services team" is used as a generalisation of all parties required in the designing, procuring, installing and commissioning of a fully operating installation. It is used for simplicity, but implies that in this modern age it is the collective responsibility for all team members to work towards a successful installation and drop the historical divided responsibilities. In areas where precise roles are required the general descriptive title of the role is given, e.g. design engineer, commissioning engineer, project manager etc.

No textbook would be able to cover each subject in sufficient depth to lead the reader to a proficiency with which they could apply the modalities to any given situation. The reference list directs readers to further sources of information. These can be used to find further practical applications of the concepts, as in the referenced documentation or deeper understanding of the underlying theories.

REFERENCES

1. BSRIA (1998) *M&E Contracting - Market Size, Structure, Sectors and Self-employment*, Bracknell, UK, BSRIA Publications.
2. BSRIA (1999) *Building Services Consultants - Market Size, Structure and Sectors*, Bracknell, UK, BSRIA Publications.

FURTHER READING

An excellent introduction to the subject of construction procurement can be found in :

3. Cox, A and Townsend, M (1998) *Strategic Procurement in Construction*, London, Thomas Telford Publishing.
4. Masterman, J.W.E (1992) *An Introduction to Building Procurement Systems*, London, E&FN Spon.

Chapter Two

Project Requirements

2.1 CLIENT REQUIREMENTS

2.1.1 Nature of Clients

The first step in determining a procurement strategy for a particular services installation is to understand the nature of the client. For building services two general categories of clients will always dominate the process: the primary contractor, who is in fact the pseudo client of a project; and the true project client, being the person or company receiving the end product. Both these client types will place differing emphasis on and express different concerns for the project. Their experience with the service installation will determine the project's success. Thus their experience with construction in general and more specifically with building services and individual members of the services team is highly significant.

During the development of a procurement strategy, four main issues must be considered: the nature of inexperienced clients, the nature of risks, the specific project characteristics and inherent attributes associated with the chosen procurement arrangement. All construction work carries risks to some degree or other. What is important in procurement is to understand the inherent risks that may be realised when the experience level of a client is mismatched with a sophisticated procurement arrangement. Non-traditional procurement arrangements may expose an inexperienced client, and consequently the services team, to greater risks. What must be established are the specific risks associated with any given project and the client's attitude to these risks. The nature of the client will determine their tolerance to certain risks.

The nature of the client is determined by two main factors: their level of direct experience with construction and the nature of their own business, particularly whether their own business operates within the construction industry environment. *Primary Clients* build for a living, e.g. government agencies, housing associations, property developers. *Secondary Clients* are more typical. They will only carry out a construction project when no other alternative is available for them, e.g. to lease more space or purchase a new building etc.

The emphasis on primary and secondary should not be understated. Primary clients building for the first time are likely to take a greater interest in the building, as a learning process for future developments. A secondary client will merely view the building as a means to an end. This difference will reflect in the participation level likely to occur.

Research has shown that 85% of clients are inexperienced secondary clients, yet the remaining 15% experienced clients account for 75% of the total expenditure within the industry. These two factors give the following general classification system that is normally used in the context of end clients.

Classification	Business Activity	Experience Level	Characteristics
A1	Secondary	Inexperienced	• Totally inexperienced often first- time client • Relies heavily on professional team
A2	Secondary	Experienced	• Experienced with an often small in-house team, but expertise brought in when required • Building does not form normal business function
B1	Primary	Inexperienced	• New developer or agency building for first time (very rare) • Builds for a living
B2	Primary	Experienced	• Builds for a living • Maximum participation due to experience with technology, design and contractual skills

Table 2.1 End Client Classifications

Whilst these definitions apply to true clients, building services are normally procured through the primary contractor, who becomes the pseudo client of the services team for the contract. The primary contractor can be viewed using the above classification in a more detailed manner. The services team should view the experience level as that held by the contractor in the particular project or technology used within the services. For example, a primary contractor constructing a hospital for the first time under a Private Finance Scheme could be viewed as a secondary inexperienced client, within that context. This will place greater requirements on the services team in terms of communication and heighten the level of risk of achieving a successful installation.

The characteristics of an inexperienced client should be seriously considered, as their nature will inevitably determine their attitudes towards and ability to accept risks. The typical characteristics of inexperienced clients are:

- lack of knowledge of project management
- lack of knowledge of industry, technology or processes
- an inability to produce a brief or prioritise requirements
- an inconsistent and ill-disciplined approach to involvement in the project
- a desire to make changes and lack of acceptance of the consequences
- being open to the undesirable influence of external parties
- being unaware of the importance of, or balance needed between, whole life cost and capital cost
- probably lacking a discipline in maintenance regimes.

These can be contrasted with the characteristics of an experienced client, who has:

- detailed knowledge of construction and its processes
- may posses in-house expertise in construction, operations and maintenance
- expertise in building services technology
- the ability to produce a comprehensive brief and balance the requirements of time, cost and quality
- a willingness to become involved with strategic decision making within the project, or use a sophisticated procurement technique
- set corporate requirements and specifications for service levels.

The public or private company status of the client is the remaining factor that determines the nature of the client. Public bodies will be constrained by their constitutions, European law, public accountability and general public acceptance of their actions. Private clients face few restrictions, other than their shareholders desires. However, increasingly private companies are becoming aware of the public presence in such matters as employee satisfaction, environmental impact and general corporate image.

2.1.2 Determining Needs

The need for a project is determined by the project business case. In the same manner, the level of services required, their technology and arrangement and the balance between whole life and capital cost will be determined by a sub-business case, specially written for the building services. And since they can account for up to 60% of the capital cost and 90% of the operating cost, building services will always have a profound effect on the business case. Along with these commercial considerations, clients must also determine their functional needs of the system. Issues such as adaptability, technology level or sophistication, life span and maintenance regime must all be considered within the business case.

Needs can be divided into the following two groups:
corporate needs: the commercial needs that dictate the viability of the project and set the commercial constraints for the project, characterised by cost, time and quality.

permissive needs: the flexible needs, often based on client values, that are related to the developing design. These are often considered as aesthetics, environmental, comfort level and image needs.

Recent research at BSRIA into customer satisfaction shows theses concern expressed under several key headings by clients on specific needs.

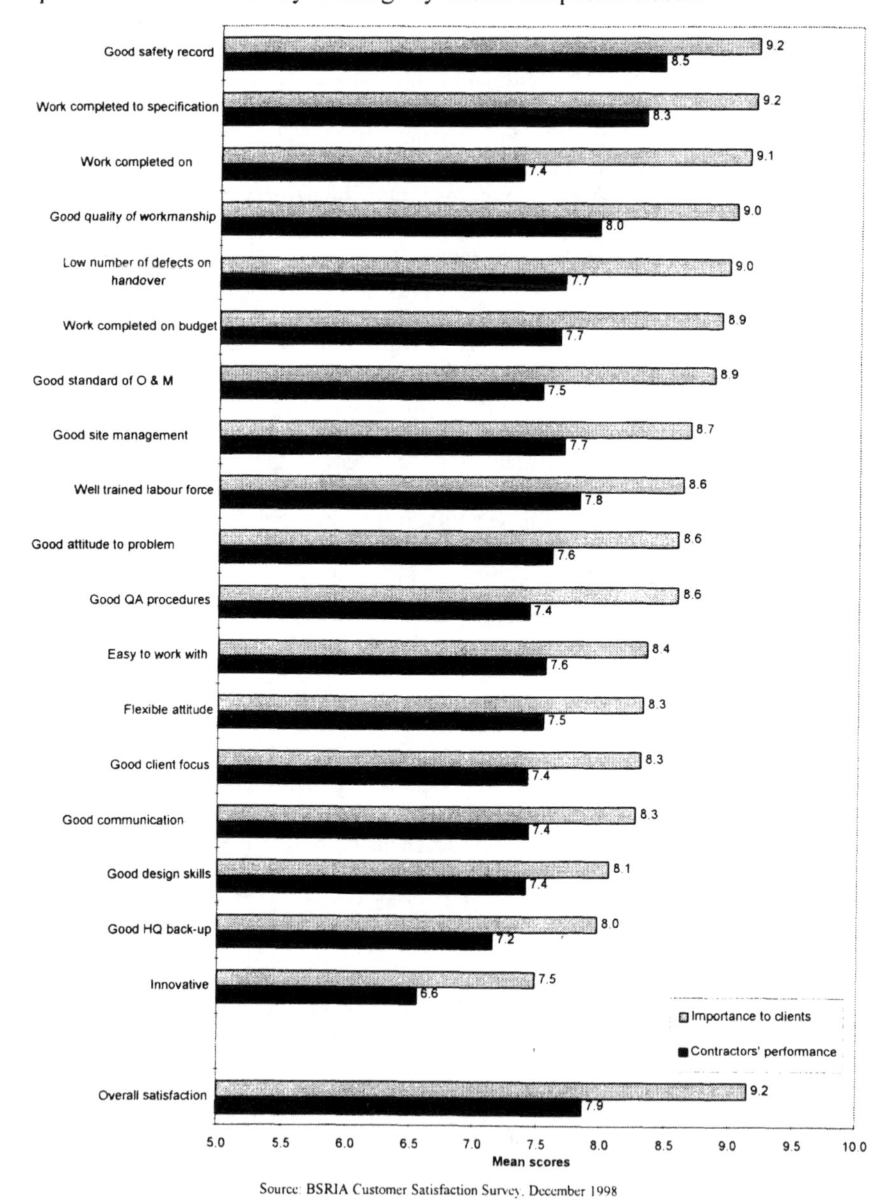

Source: BSRIA Customer Satisfaction Survey, December 1998

Figure 2.1 Clients Needs

These figures should then be contrasted against the needs expressed by clients for building services and, more importantly when selecting a contractor, their previous performance under these categories.

The initial purpose of a business case is to determine the suitability of all options available for a client to suit their need for a building. For building clients have several options, they can lease, rent, purchase, refurbish, extend their existing premises, or build new. More lately the option of outsourcing key processes, and therefore limiting their space requirements, has become another option. Only once the business case has been fully developed, with all possible options analysed, can a project move onto the briefing stage, but only if the favoured option involves construction.

This independent assessment of a client's needs highlights the first problem in procurement. It is inherent that consultants will give advice that is biased towards their level of understanding, area of expertise and need to secure work. Traditional advice given by consultants is often inconsistent, biased towards their profession and normally insufficient in providing strategic consideration for the business case for a particular need. The move towards design and construction procurement has further aggravated the problem as few contractors have the detailed skills to undertake such an analysis and suffer from the same bias as consultants.

Impartial advice can be obtained from the growing use of client advisors or principal advisors. These independent consultants provide an array of services, broadly under the heading of risk analysis, business case development, options appraisals and procurement strategy advice.

2.2 BRIEFING

2.2.1 The Business Case

The business case that is normally written for a project is a complex arrangement and an assessment of a range of factors. It is a clear statement of the financial viability of any decision. To determine clearly the best option to service a need requires an objective assessment of the overall viability. All projects, except for the very few built for personal reasons by the very wealthy, will be limited by the available finance. It is the role of the business case to anchor the client's aspirations to the overall limiting factor – money. And one of the largest factors in any whole life or capital cost decision is building services.

The important first step in developing a sound business case is to ensure the client's objectives are both well known and robust. Most clients only build to contain some form of business operation. The quality of the indoor environment, and therefore the service level required, impacts considerably on the financial viability of a building.

It is a well cited statistic that building services can account for 20% of the capital cost of an office building, and up to 75% of the cost of specialist hospitals, with operating and maintenance costs accounting for up to 90% of a building's whole life cost. Therefore a sub-business case should be developed that feeds the necessary information into the main project business case. Furthermore, the effect that the services have on building productivity needs to be evaluated. Stanhope Plc, a leading UK property developer, has calculated that over the average life span of an office building, construction costs only account for 2.5% of the total cost, yet the salaries paid to the contained staff account for 69%. It is well known that the indoor environment affects the manner in which people work and their productivity. Therefore, providing a better quality of indoor environment that affects productivity by as little as 3%, can balance the entire construction cost.

Category	Percentage	Client Business Objective
Design	0.5%	Adaptability and life cycle
Construction	2.5%	time, cost and quality
employee salaries	69.0%	Added value of staff
Energy	7.0%	energy efficiency
rent/finance	17.0%	Optimum use of floor space
Refurbishment	4.0%	minimum disruption

Table 2.2 Typical 25-Year Business Cost for a London Office Building

The sub-business case should include detailed consideration of the available capital cost, the operating and maintenance strategies that will determine whole life costs, together with a detailed assessment of the likely energy costs needed to operate the system. These costs must be considered in conjunction with procurement strategy, as the use of differing strategies will affect the cost.

Service-based strategies, where clients pay either by use or on a set user charge, can often lower capital costs and reduce financial risks, but are normally more expensive over the life cycle of a building. Detailed consideration of the building use, client's business and proposed life cycle needs to be carefully given.

Typical cost categories found within a project business case include:

- land costs
- demolition or remediation costs
- new construction cost or refurbishment of existing facilities
- operating costs
- maintenance
- decommissioning
- insurances
- design and consultancy fees
- legal fees
- in-house administration and relocation costs.

Building services are increasingly a key consideration in most of these categories. Whilst items like land cost may appear to be influenced in an abstract manner, each category is nevertheless affected by building services.

Cost Category	Influence Cost
Land	Availability and capacity of services, diversion of existing services
Demolition or remediation	Existing large-scale installations, age, type and capacity of existing services
Construction	Level and type of services provided
Operating	Services type and specification level
Maintenance	Strategy devised during design, business continuity level required
Decommissioning	Age, extent and nature of services
Insurances	Procurement arrangement and contract
Design and consultancy	Strategy adopted, together with level of service provided
Legal fees	Type used on project
Administration and relocation	Duplicating of services during move

Table 2.3 Building Services Influence on Major Cost Categories

The components of a building services business case should include detailed analysis of the capital cost, operations and maintenance costs, tax effects and any incoming services requirements. Essentially the initial business case will form the preliminary outline to a whole life cost assessment.

The final element within any business case is the risk allowance. All cost estimates will consist of a base element, plus a contingency allowance to cover the eventuality of any risk. During the progress of the project, the level of risk allowance will normally reduce, in accordance with the overall detailed analysis of both design options and the inherent risk.

2.2.2 Managing the Brief

The success of a project will largely depend on the ability of the client's objectives to be embodied within the completed building. Briefing should not be viewed as a purely front-end activity, but as a management procedure that must evolve during the life of the project, while communicating the same core message of the project's objectives. The activity continues through the phases of design, tendering and construction and is completed with a full debrief of all parties.

If briefing is about anything, it is about communication. Like value chains in the supply process, each stage of the project must take the brief and add to it. Each group must possess the key skills of listening and perception. The message must be communicated clearly and received with equal clarity.

In most textbooks the briefing stage is often shown as occurring after the appointment of the design team. This is wrong. Logical thought will quickly demonstrate that objective assessment as to the most appropriate designer to undertake a project can only occur once a certain amount of the brief has been undertaken by an independent party.

Briefing is crucial for building services, as the functionality of the final system will be determined by continuity between stages. Project requirements need to be communicated from high-end broad project strategies through to the specific tactical actions undertaken by individuals. The end success of the system will be determined by the ability of each step to be linked in the chain from client ideal to technological function.

As joined up procurement such as prime contracting and integrated arrangements gain popularity, the briefing process has become more complex. However, it is in the design process that client wants are transformed into specific requirements and thereby dictate specific supply strategies. Within these integrated procurement routes, contractors must gain the skills of good briefing practices.

Good briefing should begin with understanding the client's overall mission, being either corporate or personnel, together with the general goals the organisation wishes to achieve, and the particular objectives that a new development is seeking to achieve.

BSRIA's *Design Briefing Manual*, describes the objectives of managing the design brief as being:

- To provide a structured procedure by which the building services designer can monitor and control the briefing process;
- To provide a structured communication path between the client and the designer by which information and decisions may be passed;
- To provide a means of stimulating discussion by drawing attention to the relevant issues pertaining to the design of building services systems;
- To identify specific report stages, which will form a record of the information exchanged and decisions made.

During the first stages of a project, most clients will state their building services requirements in general and broad terms. It is paramount that concentration should be made on concepts and principles at this stage. It is only in the final stages of design that the brief should move to detailed specifics. Briefing is a translation procedure, taking general descriptions and determining specific requirements. For this reason contractors have been encouraged to become involved at the early stages of a project, as they provide a direct link between broad requirement and actual function. However, consultants usually have greater access to modelling programs, where various options can be considered.

It is important during this stage to determine precisely:

- functional requirements
- cost limits, both capital and operating
- quality levels
- expected maintenance strategies
- construction time limits

Regardless of the exact nature of a client's objectives, all will share these requirements, but will place differing emphasis on each. Taking the three traditional requirements of cost, time and quality, each client will rate these in varying percentages. The role of the consultant is to design a system which matches these set priorities. In essence, this approach is value management. This technique further attempts to link and integrate the development of a single project to the corporate strategy and operation of a client's business.

The development of the tender documents is a key stage in project briefing. For it is at this point that all of the accumulated detail of the client's requirements must be set down for others to understand and develop.

Currently tender documents are obsessed with the legal context of the project, rather than with an explanation of the goals and project objectives. Greater project success can be accomplished by a non-adversarial approach at the tender stage. By clearly stating the requirements in an open forum, problems on both sides can be pre-empted, evaluated and eliminated.

By taking a proactive approach during the tender stage, the brief becomes a teamwork document. All parties commence the project with a full understanding of what is required and more importantly, how to work together to obtain this common goal.

As stated previously, good briefing revolves around good communication, which can be broken down into three main constituents:

- direct communication and relationships between participants
- management of the process
- information used for the briefing.

During each stage of briefing these three elements are used to communicate the client's mission. Interference at each stage of the message and the relay of information may occur. Current trends in procurement have attempted to minimise this interference by ensuring that the contractors are involved early enough to participate in the development of the detailed brief.

This arrangement provides a workable response to the main constituents of briefing. The information produced by the design team will be directly compatible with the contractor's needs; a communication is unambiguous between client and provider.

2.2.3 Determining the Brief

A brief should always commence with a clear statement of the client's objectives. These must then be translated into the cornerstones of technical characteristics, performance criteria and quality standards, and be underpinned by the constraints of the available budget and time. This will enable all parties to remain clearly focused on the end goal. Briefing is as much to do with psychology as it is with technology.

Most clients will not posses the detailed technical understanding of building services, and therefore will normally describe their requirements in vague and general terms. This vagueness and inexperience contrasts against that of the building services engineer, who is steeped in technology, but has little knowledge of the client's business requirements. By not admitting their weaknesses in understanding, problems can develop. Inexperienced clients will too readily agree with an assertion made by the engineer, rather than debate their true requirement. Similarly the consultant must constantly probe as to nature of their business and its true requirements.

The brief should be a free-flowing dialogue between client and the building services team. Only within a relaxed atmosphere of trust will clients disclose their needs. This will take considerable time to develop and consideration should be given to varying both the location and format of briefing meetings. Quite often informal walks around unrelated buildings to those of the clients, will reveal personal views.

This perhaps raises the issue of using a client's advisor, who is divorced from the design team. But although a client advisor should be able to pass on a properly considered brief, the interaction with the client is lost, with the brief becoming a single-sided statement of needs. Clear assessment must be made during the development of a procurement strategy as to the interaction required with the brief. Each procurement arrangement allows differing levels of communication and feedback with the client. The long-term role of the client advisor will also need to be considered once the initial briefing stage has been completed.

Complexity is added at this stage, as the client is rarely singular, but is a team. Each party in the client team will add to the debate and each will possess a differing level of knowledge and ability to communicate it. Ideally the client should assemble a briefing team, representing the key functions within the business. The team should include:

- facilities manager
- office manager
- operations director
- health and safety officer
- finance director
- maintenance manager
- employee user groups.

A Johari window is often used to explain the concept of knowledge disclosure between the client and the advisor. The basic idea is to release as much information as possible to the other party, causing the unknown window to become as small as possible (see Figure 2.2). Public information represents the information that is clearly communicated by the client. Blind information is known by the advisor, but the client is unaware of it or its importance. Clear communication of these needs can lessen the amount of blind information. Likewise, private information is known by the client but is not relinquished, purposely or otherwise. Unknown information is that which is not understood by either party and needs to be minimised. Such failure to be fully aware of all relevant facts could result in changes having to be made during both design and construction.

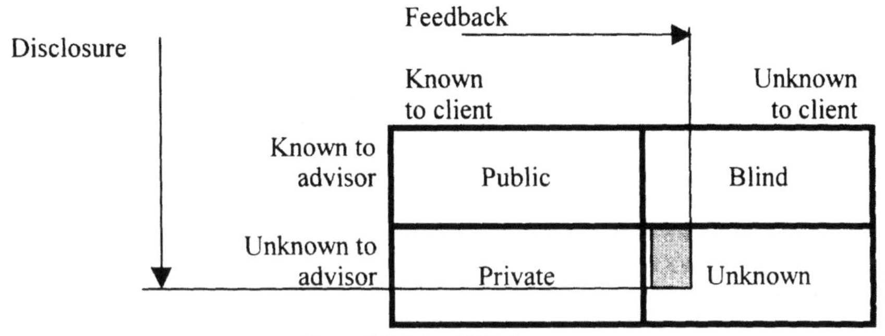

Figure 2.2 Johari Window

So the briefing process must remain fluid and based on clear communication, rather than being hemmed in by detailed procedures. A project brief should not be a miscellaneous collection of client's wishes and desires, but a structured document that breaks each objective down into the required facilities and services needed to achieve these objectives.

The brief will develop and change through the project life cycle. Four key stages exist, each requiring greater detail as the project progresses.

Design stage	Consideration	Brief Type
Inception	Outline requirements and future actions	Inception Brief
Feasibility	Options analysis and selection of most suitable functionality, technically and financially	Strategic Brief
Outline Proposal	General approach to layout, design and construction, checked against cost, time and quality constraints	Project Brief
Scheme Design	Detailed requirements of location, output, size, and co-ordination of outlets with architectural requirements	Consolidated Brief

Table 2.4 Design Stages

Procurement and briefing can often be at odds with one another. In fact the idea of clearly establishing both creates a paradox. The procurement strategy cannot be properly developed until the precise nature and extent of work is known, yet this requires the input of professionals, who cannot be involved until the procurement strategy is developed and executed!

2.2.4 Briefing Stages

Inception Brief

Also known as a client's brief, this document can best be viewed as a self-discover exercise on behalf of the client. The document is a series of statements made by t client that sets out the need for the project. The exact content will depend upon the project nature, the client's abilities and experience and the advice they have already obtained.

As a minimum the brief should contain:

- the client's business function
- the client's business objectives
- the structure of the client's business or organisation
- the client's perceived need
- any relevant background information
- the consequences of failure or any other perceived risks
- the nature of advice needed to progress the project.

Most importantly the document should contain a statement on the perceived procurement requirements, including accountability, risk tolerance, competition requirements and required delivery schedule. Assessment of this document by a principal advisor should form the basis of any further development required, based on a preliminary assessment of possible procurement options.

Strategic Brief

Although dependent upon the client's ability, the strategic brief is normally developed by the principal advisor. The purpose of the brief is to agree the basis for feasibility studies and engineering reports that will be fed into the assessments regarding alternative design options.

The strategic brief should contain:

- the mission statement of the project
- the scope and purpose of the project
- functions and accommodation to be provided for
- the client's objectives, with relative priorities stated
- capital expenditure limits
- targets and constraints
- quality requirements
- scope for further studies, surveys, etc.

From the strategic brief, a project execution plan should also be written, which sets out the major commercial and procurement requirements for the project. It also sets out the headlines for the development of a project procurement strategy:

- proposed procurement strategy
- organisational structure of project team
- communication plan
- health and safety plan
- quality objectives

- risk management objectives
- value management proposals
- environmental policy
- budget and cash flow forecasts
- project master programme.

Project Brief

The project brief forms the definitive set of requirements for the project, allowing the design team to begin the detailed design. Depending upon the procurement strategy adopted, the detailed brief would normally be completed by the design team leader, in conjunction with the briefing committee.

The brief should state all necessary technical, managerial, commercial and design objectives that were identified within the strategic brief. Table 2.5 below includes the issues specific to procurement. Often in building briefing issues such as facilities management, function and risk levels (for systems failure) are ignored. Operational issues must also be clearly established to ensure the level of technology adopted for controls systems and delivery of the building services function is in line with the clients ability and overall operational strategy.

General Description	
▪ site location and features	▪ environment /weather factors
Project Objectives	
▪ user needs	▪ business objectives
Project Brief	
▪ purpose/function of the project	▪ maintenance requirements
▪ accommodation schedule	▪ business criticality demands
▪ quality standards	▪ environmental factors
▪ operational requirements	▪ disposal criteria
▪ specialist services/equipment	▪ statutory requirements
▪ occupancy levels and patterns	▪ building/services life span
Constraints	
project	*corporate*
▪ planning conditions	▪ company policies / standards
▪ listed building status	▪ design standards
▪ utilities availability	▪ undertakings given
Controls	
▪ financial budget	▪ quality standards
▪ timescales and milestones	▪ cash flow
▪ acceptable risk level	
Prioritisation	
▪ cost vs. time	▪ time vs. quality
▪ quality vs. cost	
Occupation	
▪ facilities management	▪ commissioning
▪ maintenance	▪ operation
▪ handover	▪ timescale

Table 2.5 Project Briefing Headlines

Consolidated Brief
Once the detailed design has begun and most design options have been determined, then the detailed design brief should be presented to the client as a consolidated design brief.

It should contain the following information:

- a list of project principals
- summaries of option studies
- a summary of design discussions and reasons for design choices
- a description of adopted design solutions
- a construction cost plan
- a design programme
- outline design drawings
- supporting design calculations.

The brief will form the benchmarks and parameters for the project. It must precisely reflect the requirements and be fully agreed with the client before proceeding with the detailed design.

2.2.5 Debriefing

The debriefing stage is crucial, as it allows all parties to obtain feedback on their performance. By accepting it as constructive criticism, performances can be improved.

Typical benefits obtained from debriefing include:

- assisting suppliers and contractors to deliver more successful projects
- gaining mutual feedback of procedures, marketplace, system design and performance
- increasing mutual opportunities for value for money on future projects.

All clients should undertake to provide detailed feedback on unsuccessful tenders. Government clients are legally obliged to provide this to contractors. Under EC Regulations, government must provide specific reasons for not selecting contractors, together with details of the successful party.

Business benefits can be gained from tender feedback by:

- encouraging suppliers to submit further bids in accordance with the lessons learned
- assisting suppliers and buyers to improve performance and procedures
- demonstrating interest in suppliers' proposals and business, thus educating both parties on each other's future requirements
- increasing competition in future bids.

Project debriefing should be undertaken as a lessons learned exercise. Problems identified, whether during construction or design, concerning logistics or people, should be recorded discussed, and used as a benchmark for increasing value and performance.

2.3 ROLE OF THE MAIN CONTRACTOR

2.3.1 Project Management

The Main Contractor

As the move towards integrated procurement arrangements continues, where a single contractor undertakes the overall management and determination of requirements, the role of the main contractor continues to evolve. Projects are now being delivered as a composite product, requiring the traditional "mailbox" mentality of main contractors to change. Increasing sophistication of management techniques, risk analysis and technology have also played their part in accelerating the evolutionary process.

Since the Second World War, main contractors have evolved from being the holders of all knowledge and resources to complete a project, into the now divided camps of:

1. **professional contractor:** characterised by contractors that undertake work normally under a management only regime, such as that characterised by construction management, management contractor or similar procurement arrangements.
2. **general contractor**: the traditional role of the main contractor evolved to where limited resources are inputted, typically for general labouring, with all work sub-contracted out under a domestic arrangement. Their role is that of strategic co-ordination of these sub-contractors. Traditionally, this relationship has been compromised due to the adversarial nature of the contractor.

It is increasingly common for large contractors to undertake both roles, but with often separate specialist divisions.

The development of architectural styles and services technology during the 1960s caused the birth of specialist contractors. Curtain walling, air conditioning, control systems, structural steelwork and alarm systems were typical developments in this area. These technical complexities fuelled management complexities.

As the complexity of technologies increased, more specialised contractors and consultants developed. The result is an industry divided by a common language. Each specialism developed its own methods and language, increasing the need for, and causing problems in executing, proper co-ordination and integration of the specialist technologies.

A disparity began to exist between the knowledge held by specialist contractors and consultants, and the general knowledge of the main contractor and principal consultants, namely architects and quantity surveyors. Little is written about the impact of these differences in knowledge, which has caused the well-publicised adversarial environment. As the main contractor's knowledge became sidelined, they sought stricter controls through contracts, rather than developing their understanding.

Along with general management duties, the main contractor also normally undertakes the prescribed legal duties, as defined under the Construction (Design and Management) Regulations 1994.

The Regulations require the appointment of a principal contractor, who under the specifics of the Regulations, must be a contractor, although the named person or company can fulfil the dual roles of planning supervisor and principal contractor. However, it is the client's responsibility to ensure both individuals, and all other contractors and consultants, are competent prior to their appointment.

Although the Regulations dictate actions that must be undertaken by the principal contractor, all of these are of basic common sense that should normally be complied with under good management practice. Basically, the principal contractor must ensure that:

- all actions of contractors on site are properly co-ordinated and comply with the health and safety requirements
- all appointed contractors are competent
- sufficient health and safety provisions have been made within a contractors work plan
- only authorised people enter the construction site and are briefed on the dangers and safety requirements.

A contractor cannot allow his employees, or sub-contractors, on site until they have been provided with the following information:

- the name and details of the planning supervisor and principal contractor
- the health and safety plan, or the parts relevant to the contractor's section of work.

Contractors must comply with the directives of the principal contractor and those imposed through their statutory duties. Furthermore, the Management of Health and Safety at Work Regulations 1992 require contractors to supply the principal contractor with all relevant health and safety plans, risk assessments and method statements required for the safe execution of the work. The contractor is then bound by these documents, together with the project health and safety plan.

The Act of Project Management

Given the move towards product-based procurement strategies, where the main contractor is responsible for the design, development and construction of a project to the stated requirements, the main contractor thus takes on the role of professional project manager. This singular project role must be balanced with the long-term strategy of the contractor's business, which often results in a compromise.

To be successful the main contractor must deliver a project in line with the clients objectives, yet simultaneously service their own corporate needs. Corporate requirements are often general and measured over an extended time period. Project management has a defined life span and set goals, often described as cost, time and quality objectives.

These objectives are executed through the general project stages of:

- conceiving and defining the project
- planning the project
- implementing the plan
- completing and evaluating the project.

Each participating organisation will also require project management to achieve their sub-project goals. Generically, sub-project managers would include the client's representative, design team leader, construction manager and the specialist contractor.

....

Case Study

The American Perspective

Much has been written about the differences between UK and US construction techniques. Although most have concluded with startling figures of up to 30% cost savings, the real truth is the way both countries approach construction is so radically different it defies an easy comparison.

One of the leading reports, The UK Construction Challenge published by BAA's then property arm Lynton studied the construction of identical buildings built at Heathrow UK and Charlottesville North Carolina. Its conclusion was that construction costs were comparable when identical building specifications were used, but a saving exceeding 30% was apparent when the Charlottesville project used a "standard" specification. The majority of the saving was attributable to the use of an off-the-shelf air conditioning unit.

The report concluded that the pre-construction phase was significantly different and observed the following issues:

- In the US, there is a greater integration of cost and design. Cost management is not a separate specialist skill, and design is developed within strict cost limits.
- Single point design responsibility. In the US, architects are typically responsible for the entire design and engage engineers as sub-consultants.
- Greater standardisation of trade packaging in the US. Work is procured on a trade basis that is common from project to project.
- In the US there is earlier technical and cost input from trade contractors, with innovation expected and encouraged.

In the US, construction managers are appointed earlier, have greater status and authority and possess stronger technical skills.

REFERENCES

1. BSRIA (1998) *Customer Satisfaction Survey*, Bracknell, BSRIA Publications.
2. CIBSE (2000) *Guide to Ownership, Operation and Maintenance of Building Services*, London, CIBSE Publications.
3. Darlow, C. (1994) *Valuation and Development Appraisal*, London, The Estates Gazette Limited.
4. Lynton PLC (1995) *The UK Construction Challenge*, London, Lynton PLC Company Report.
5. Masterman, J.W.E. (1992) *An Introduction to Building Procurement Systems*, London, E&FN Spon.
6. Parsloe, C. (1990) *A Design Briefing Manual,* Bracknell, BSRIA Publications.
7. Turner, A. (1990) *Building Procurement*, Basingstoke, The Macmillan Press Ltd.
8. Wild, J. (1997) *Site Management of Building Services Contractors*, London, E&FN Spon.

Chapter Three

Contract Strategy

3.1 CONTRACT RISK

All construction projects involve risk. The traditional method in which risk was dealt with during procurement was through the contract, which led to many disputes and the general view that construction is an adversarial business. Contracts have now become politically incorrect, being synonymous with the lack of strategic decision making in procurement.

Traditionally, contracts were used to reinforce poor decision making or injudicious risk transfer. This common view is often associated with contracts that were largely written by biased professional parties, with the contractor poorly represented. The contracts served neither the parties responsible for the risk, nor the client's strategic intentions of the project. However, contracts are a state of mind. All contracts, regardless of their simplicity, could be viewed as adversarial. What will truly determine success is the manner in which they are executed and managed.

Main project contracts are now being redrawn to rebalance the responsibility between the various members of the construction team. Depending on the actual main project procurement strategy, the building services team may work under one of these standard contracts, but is more likely to be subject to the internal conditions set by the principal contractor.

Regardless of specifics, the three key aspects of a contract must always be to make a clear statement of responsibilities and to appropriately allocate risk, together with setting the work within the proper context of the clients objectives.

This chapter is not intended to be a complete guide to the selection of specific contracts. The modalities included within Chapter 4 have been developed to assist in the selection of the more common forms of sub-contract agreement. The main purpose of the chapter is to introduce the subjects that must be considered during procurement, especially that of understanding risk.

3.1.1 Nature of Risk in Procurement

All contracts stem from the nature of the project and therefore from the decisions made by the client. To understand how risk will affect the project and how it is best handled through the use of contracts, the interrelationship between the client's personality, procurement strategy and the nature of risk must be understood.

Regardless of the client's nature or characteristics, the main criteria for project success, and therefore their level of participation within the procurement process, determines the eventual level of risk endured by the client. The technical complexity of the project remains the same regardless of its position within the world, therefore the only variable is within the client strategy, in terms of the objectives and nature of the client.

It could be concluded that it is not the procurement arrangement or the level of risk that will determine project success, but rather it is the level of participation from the client. However, this assertion does not recognise the fact that the traditional system deliberately distances the various parties and minimises the involvement of the client. It is important for clients to recognise their boundaries or the limitations of their knowledge.

Clients become involved with a project in order to;

- state clear objectives
- be unequivocal in decision making
- build a strong and effective design team
- ensure allocation of risk to the appropriate party
- take responsibility for decisions
- create a co-operative problem solving atmosphere
- maintain accountability
- maintain control.

To clearly state objectives and maintain control of the project requires detailed understanding of the technology involved. And for building services few clients possess such knowledge. With this contradiction the emphasis must be further stated that inexperienced clients require guidance, preferably from an unbiased source.

Risks that directly affect clients during the procurement of a building project stem from one of the three main areas:

- financial risks - associated with both the capital cost of the installation and its subsequent running costs,
- construction risks - associated with the building, installation and commissioning of the building,
- design risks - associated with the success of the design meeting the user's requirements.

The influence these risks have on the client varies with the procurement strategy that is developed for the services installation. However, the financial risk of any venture will ultimately lie with the initiator and cannot be discharged, unless a strategy using service provision, such as when PFI (Private Finance Initiative) is adopted. Obviously any party participating in the building services team will run the financial risk of carrying out the work for the stated price, including the risk of the client actually paying!

The effect the other risks have on the client can either be minimised or eliminated. With any risk passed to another party a premium will be charged against the client for the acceptance of this risk. By offsetting the risk to another party the client may appear to be minimising their risk but may increase the risk of claims or litigation from a contractor (this is known as secondary risk). No contractor can take a risk for which they cannot be financially compensated.

The risks associated with construction are concerned with the four major principles of the contract: quality, quantity, time and cost. Although normally considered to be construction risks, the same headings can be applied to design. In fact the chosen design solution will have a major impact on the level of risk required during construction. This gives the client a choice with procurement arrangements: either separate design and construction and run the risk of the two principles clashing (as in the traditional arrangement) or choose an arrangement in which the two main principles are combined (as in the integrated arrangements now being advocated).

The consequence of any risk is either direct or indirect. Direct risk will affect the client outright, such as late completion, while indirect risk might occur through claims by the contractor or higher tender prices. By offsetting risks with the chosen procurement arrangement, the client may be avoiding direct risk but increase the risk of consequences from its offset.

Construction is the execution of the works prescribed by the client. By selecting the appropriate procurement method the client is able to alleviate themselves of the construction risks. Most contractors deal with the inherent risks of their trades, either through pure management or minimising their impact through standard insurance policies. However, the risks that pass to the contractors must be valued and within their ability to control. The rights and obligations of the contract must match the risks being accepted by the contractor. By transferring too great a risk within a contract, the client suffers either from high prices or through possible legal consequences.

3.1.2 Assessing Risk

'No construction project is risk free. Risk can be managed, minimised, shared, transferred or accepted. It cannot be ignored' *Constructing the Team. Sir Michael Latham*

Although risk is normally associated with adverse happenings, upside prediction - when positive benefits can be incurred from a risk, as in the stock market - in business forecasting plays a major part in business planning. Understanding the basic principles allows the application of risk management within everyday decision making.

It is particularly apt for building services as the failure of the designed system can have devastating clerical and human consequences. Sophisticated computer modelling will eliminate these events. By analysing the probability of failure, product design and installation can be modified to decrease the probability of failure. In the same fashion risk management used in whole life costing can be used to predict significant savings in alternative maintenance strategies.

Due to the uniqueness of the industry, risk management needs to be proactive. Although all construction projects share the same components, their unique locations and geographical conditions makes each susceptible to increased risk. Setting clear objectives for any project is a fundamental requirement of any strategy. Poorly managed risk will adversely affect the ability to achieve these objectives. Like all decision making, dramatic changes can be made in the early stages without incurring cost penalties. The initial options appraisal of the intended project may have done so utilising all known or possible risks. The developed risk register will continue through the project, updated as and when risks are identified.

The principle behind risk management systems to measure all risks are identified and understood. Those retained must only be those for which an appropriate management system exists for their control, together with adequate contingencies. The end objective must be to ensure that the project objectives are met and value for money is delivered.

It must be first understood that risk management is not about trying to predict the impossible or even the future, both of those are called uncertainties. Risk management is about quantifying the outcome of alternative decisions, ensuring the decisions being made today will provide a satisfactory basis for decisions tomorrow. Business forecasting works in a similar manner allowing commercial decisions to be made to increase financial results. Excessive concentration on the negative leads to conservative and irrational decision making. Taking measured risks is required in all aspects of business. Growth and business development all require decisions on risk and uncertainty.

Precise definitions of these concepts can be expressed as:

- Uncertainty: an event where the outcome or its consequences cannot be foreseen. Usually termed force majeure in insurance policies.
- Risk: a quantifiable measurement of the possibility of an event occurring multiplied by the magnitude of the loss or gain. Risk differs from uncertainty in that it can be quantified and therefore insured against.
- Business Forecasting: the use of risk based on quantitative forecasting methods to determine commercial gains or losses.

To differentiate risks from consequences, two parts must exist: probability and impacts.

This gives four generic risk types:

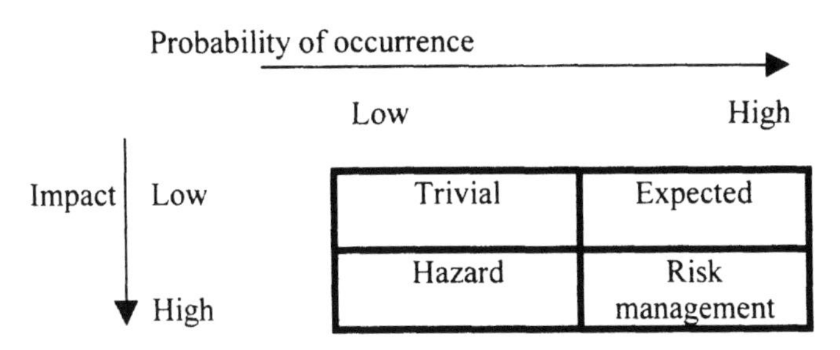

Figure 3.1 Generic Risk Types

The process begins at the earliest possible stages of a project. Experienced developers use scenario planning based on various factors from interest rates and void periods, through to rates of return, to determine the suitability of developing a project. Although sounding complex it normally takes the form of brainstorming sessions with various advisors to identify as many risks as possible. For issues such as interest rates a sensitivity analysis gives a spectrum of results for analysis.

Risk measurement and analysis techniques have a wider range of possible applications than just risk. They are widely used in the examination of decision making in property investment and development strategies. Used within business forecasting they can enhance the reliability of estimating installation costs, whole life costs, finance rates or the future economic performance of business units.

Risk management begins with the analysis of the project, to determine specific risks and their sources. Impacts on time, cost or human consequences can then be measured. An appropriate response to the risk can then be given, ensuring contingency plans exist to deal with retained risks. The final stage is for an appropriate feedback system to be implemented, allowing gained knowledge to be used in further identification responses to risks. Properly executed, the management system allows managers to concentrate on the critical success factors of a project.

Internal Project	External Market
▪ complexity	▪ inflation
▪ speed of design	▪ market conditions
▪ location	▪ cost escalation of resources
▪ size	▪ labour / materials availability
▪ novelty	▪ political uncertainty
▪ construction methodology	▪ weather

Table 3.1 Major Sources of Risk

The benefits of using risk management occur as a result of the explicit highlighting, analysis and evaluation of issues that are fundamental to a project. Difficulties in using the techniques extend from problems in predicting future events, conditions and outcomes. These can be counteracted by using adequate and reliable data.

However,·the fundamental problem with prediction techniques is that they rely on the belief that the past presents a suitable model for the future.

A risk analysis is a formal review of retained risks. Its main objectives are to:

- identify retained risks
- identify new risks as the project is developed
- assess any unusual characteristics of the risk
- determine maximum and mostly likely outcomes.

Risk and contract strategy are interrelated, as is the procurement strategy. Selecting the appropriate contract will allow the risk response strategy to be implemented and understood by all parties. Standard contracts should always be preferred, as their principles and robustness have been tried through previous court rulings. All contracts allow for procurement strategies to allocate risk to the various contract parties in different manners. With the allocation of risk goes the authority and responsibility for controlling it, as does the right to seek compensation. Transferring risk through a contract is fair as long as it is understood and all parties are allowed the opportunity to price for its responsibility.

Main contract procurement strategies have recently been developed for maximum transfer of risk to the contractor. Government believes that private industry is more equipped and knowledgeable of risk, and therefore more commercially effective in its management. Coupled with partnering agreements, contractors are given the opportunity to fully understand the risks involved and price for them. The MOD's prime contracting rote requires pricing of life cycle costs with assurances of their accuracy. Risks of the capital costs and running costs are fully transferred to the private sector. Value management is introduced to ensure these are realised and the most cost-effective solution for all parties is developed.

Risk management and analysis should be viewed as a dynamic tool within the overall management of a project. Used effectively it allows the end objective to be concentrated upon and realised. The subject is extremely complex and full justice to it cannot be done within a single short section. Qualitative methods are simple in both execution and technique. They are suitable for any stages of the project, either commercial or technical. Understanding their application would enhance any managerial decision.

....

Case Study

Water supply and storage has increasingly been an area where detailed risk assessments have been carried out. This has been caused by an increase in awareness of issues such as legionella, cryptosporidium and other microbiological hazards.

Building owners are particularly aware of these issues, as under the Health and Safety at Work Act 1974 legal liability for the consequences of water hygiene cannot be transferred to a contractor. Legionellosis and water hygiene risk assessments are mandatory in most non-domestic buildings, particularly those with wet cooling towers or evaporative condensers. Like all risk management systems, the Act requires the compilation and maintenance of a risk register, including updates and when alterations are made to the system.

Two responses are available to limit the risk of legionella: firstly, retain by using control methods such as water treatment regimes that eradicate or control the proliferation of the bacteria; secondly, prevention by eliminating the possibility of exposure by the substituting of wet cooling towers for dry ones.

Water hygiene risk assessment involves the examination of microbial contamination of water systems. Since contamination is dependent on the environmental conditions (oxygen, light, scale etc.) and the availability of nutrients, risk management must depend on engineering a system that minimises the risk of microbial growth factors and prevent organisms passing through the system to make human contact. The use of chemicals could prevent one risk, but raise another through chemical hazards. It is impossible to eliminate organism risk, but risk management should be used to minimise the risk through good management and maintenance procedures.

....

3.1.3 Allocating Risk

The procurement strategy sets out the general relationships, roles and responsibilities of individuals within a project. When considering a strategy the nature of risk must be considered in conjunction with each party's role, and their generally accepted responsibility, together with the overall ability of a company, (both financially and managerially) to deal with a risks realisation. Risk can only be allocated to a company or individual that is capable of dealing with it.

Transferring a risk to an inappropriate party will eventually lead to an increase in the cost of the risk, thereby causing a higher secondary risk in addition to the first.

The first step in allocating risk is to clearly identify all risks that the project faces, regardless of their nature or source. This is best done through the use of a risk register. The ability to identify all risks is dependent on the team's experience. Consideration should be given to the use of either workshops, facilitated by a neutral party, where all members of the project can contribute, or the use of a professional risk consultant. The modalities in Chapter 4 provide a pro-forma risk register.

Once identified, the main instrument used to allocate risk is the contract. Historically problems have occurred by the use of adversarial contracts which have apportioned risk to the lowest level in the contractual chain, namely the specialist contractors. Too often unacceptable financial risk and responsibilities were given to companies that had neither the experience nor financial resources to deal with them. The result was risk dumping rather than a strategic approach to its allocation.

In the development of a procurement strategy, where the allocation of risks is strategically decided, five general options are available. Consideration of these options must be undertaken in conjunction with the contract strategy that is used to implement defined responsibilities.

1. **Transfer:** through subcontracting, insurance or delegation, transferring the risk to someone else with greater ability or knowledge minimises commercial consequences. Risks should only ever be transferred to a party that is better placed to exercise more effective control over it.
2. **Reduce:** minimisation of outcomes should be an immediate response, whether used singularly or prior to transferring or retaining.
3. **Retain:** once a risk is identified a positive management system can be implemented to ensure that its realisation is kept to a minimum or adequate contingencies are in place. An identified problem is a risk rather than an uncertainty.
4. **Eliminate:** the preferred option. Eliminating risk through either controls or design is always preferred, but may cause new or consequential risks to be realised.
5. **Shared:** retained risks can be jointly shared. Characterised by joint venture agreements, these normally occur when a retained risk is too large for a single company. For a share of the rewards partnerships are formed, but are equally responsible for joint and severally liable consequences of a risk eventuating.

To properly allocate risk, the project must be precisely defined to allow for some arrangements. Traditional lump sum arrangements do not allow much flexibility for change, but afford greater transfer of risk to the contractor. Management-based arrangements normally allow more innovative design solutions to mid-project changes, due to the involvement of the specialist contractors. Consideration of these responses should be made with reference to the following questions about the procurement strategy:

- How realistically is the project defined and are amendments likely during construction that may affect the risk response?
- What trade-off are you willing to accept between desired changes in design and compromises in actual execution?
- Can the project be precisely described in terms of specialisation, performance requirements and quality standards?
- Are the risks involved complex and do they require handling by specialist parties? e.g. asbestos, radiation, underwater work?
- Is there a ready source of contractors capable of dealing with such a project?
- What effect will the eventuation of risks have on the stated cost, time and functionality of the project?

3.2 NATURE OF CONTRACTS

Regardless of the conviviality that may exist between contracting parties, a formal contract that clearly sets down the obligations of each party is always required. The arrangements and responsibilities that are undertaken to deliver a fully working services installation are complex and some form of written agreement is required to clearly state the roles and responsibilities of each party required to achieve the end goal.

The formal contract consists of two major parts. Firstly, it sets out the contractual terms which define the legal relationship between the parties. These terms specify the responsibilities, liabilities and apportionment of risk to each party. Secondly, it holds together the various other documents, such as drawings and specifications, which state the deliverables of the contract.

Most contracts will comprise the following:

- agreement between parties
- conditions of contract
- consideration to be paid
- formal tender documents
- specification of work
- schedule of work (if applicable)
- drawings (if applicable).

For any given project, there are various aspects of the project that may have separate contracts, all of which will vary in length and type. The most common aspects of a project that will have separate contracts for work include:

- precontract work (either design or site preparation/remediation work)
- design and engineering
- main project construction
- specialist engineering work
- major materials supply
- commissioning/handover
- maintenance/operation.

The extent to which the project is separated into these contracts will depend upon the procurement strategy and the level of involvement and amount of control desired by the client. Certain main contract procurement strategies such as construction management may involve several contracts for each element. Conversely PFI (Private Finance Initiative) or turnkey type arrangements would normally involve only a single contract with one organisation.

In deciding the contract strategy, the client will have to consider the manner in which a contract deals with the following issues:

- timing
- manner in which variations are to be handled
- quality level
- price certainty
- apportionment of risk
- project complexity
- management style.

For most projects three contracts will dominate the strategy, each representing the major elements of the project: the main contract for the overall management of the project; consultant contracts for the design and engineering work; and the subcontract agreements for the specialist operations and work execution.

3.2.1 Nature of Subcontracts

In the eyes of the specialist contractor only two relationships are evident between various contracts, despite the proliferation of contracts and procurement arrangements over the past decade. Specialist contractors, and now most consultants as well, either contract directly with a client (as in construction management) or with the primary contractor (in all other arrangements). Therefore the final performance of the project will be dictated by the subcontract agreements between specialist contractor and principal contractor.

This simple fact is often forgotten when contractual strategies are developed. In the development of a procurement strategy for building services careful consideration needs to be given to the manner in which each works package will be procured, the contract to be used and how each of these will form into an integrated contractual chain.

Setting the main contract aside, four principle methods exist in which to form a contractual relationship with a specialist engineering contractor: direct, nomination, domestic subcontractor, and naming.

For most of these, standard forms of contract exist. However, as the contract is between the principal contractor and specialist contractor, the contract is usually heavily modified in the principal's favour. Common amendments usually include extended payment periods, additional risk transfer and convoluted appeal processes. Previously, legislation such as the Housing Grants, Construction and Regeneration Act 1996, has been enacted within the UK to prevent such abuses. Despite this, heavily modified contracts still dominate.

More importantly, within the context of building services, none of the standard forms of contract fully respect or understand the complex interrelationship required for a successful installation. The major drawbacks of these standard forms of contract when dealing with building services include:

- lack of interconnectivity between the various contracted parties
- lack of defined design responsibility
- lack of provision for commissioning and /or testing
- lack of provision for contractor's design portion
- lack of provision for detailed costing other than capital cost
- lack of provision for maintenance or system operation.

This lack of standardisation, together with the dominance of imposed conditions set by the main contractor, can cause severe hardship for the client in the event of a problem arising with the services installation. The chain of liability does not allow a direct contractual relationship between client and subcontractor. Therefore the only redress for the client is through the main contractor. This singular situation highlights why both the procurement and contractual strategy must be well thought out and properly executed.

3.2.2 Appointments

Subcontracts within building services vary considerably depending upon the nature of work to be undertaken and the extent to which a physical entity is produced. These will vary from labour-only works, such as contracts with self-employed operatives, through to comprehensive design, engineering and construct contracts for sophisticated elements of work. Between these two extremes lies a myriad of possibilities including material-only supply and expert services such as commissioning. Furthermore, the location in which the work may be undertaken can also vary. Although work is normally considered to be executed on the specific project site, the wider use of prefabrication means that a considerable amount of work can take place in a variety of locations.

The manner in which appointments are to be made should be determined within the procurement strategy. The decision must be made in a manner that supports the strategy, while ensuring sufficient mechanisms exist to allow the project to be properly managed. Specific issues that must be resolved within the procurement strategy that will inevitably determine the appointment method include:

- responsibility over the appointment
- the extent of client involvement
- the extent of consultant involvement
- the required timing of the appointment (nomination allows early design involvement)
- the technology involved
- direct bearing of subcontractors appointment on project success
- the required design input.

The final issue above - design input - can often be the overwhelming factor. Design within building services is a highly convoluted process requiring nearly each specialist to have some participation in the design. The required interaction between designer and specialist needs to be considered, yet should not compromise the commercial perspective of the project.

Each method of appointment includes advantages and disadvantages, all of which are largely dependent on the specific contract used. These can now be reviewed in accordance with dominant contract for each method. It should be noted that what follows is a general overview of the each method. It is not intended to be a detailed legal discussion, as each arrangement is subject to the terms and conditions of the specific contract. The reader is advised to consult one of the quality legal texts available that provides in depth discussion on the subject.

Domestic

This type of work is exactly what its title suggests: domestic. It is a contractual relationship only between the principal contractor and the subcontractor. Neither the consultants nor the client can influence the appointment or the conditions of the contract, although most main contracts do carry provisions for the approval of domestic contractors by the client.

Although standard forms for domestic subcontract agreements exist, it is common for these to be either heavily modified or for the principal contractor to impose their own conditions. The greatest difference between domestic and other types of appointed subcontractors is that the terms and conditions they work under are negotiated singly with the principal contractor. Therefore they are excluded from any responsibility or involvement with the project team or even any coherence with the conditions of other domestic subcontractors. This can cause problems when developing a procurement strategy as each domestic subcontractor could be working under different contract conditions, with differing timescales, payment methods, incentives and penalties.

Good procurement practice dictates that standardised contracts should always be used, as considerable time has been spent to ensure the interconnectivity between their constituent parts, together with sufficient case law existing to highlight their robustness. It is important that the obligations set by the employer within the main contract are stepped down to the subcontractor. This will also provide the subcontractor with the same rights and privileges enjoyed by the other parties. To work effectively the two agreements must be complementary.

Named

To overcome the tied legal responsibility of the nominated subcontractor (see explanation below) to the client, the JCT IFC 84 (Intermediate Form of Contract produced in 1984 by the Joint Contracts Tribunal) contract introduced the idea of named subcontractors. The idea was that clients could name a subcontractor that the principal contractors could use, but be devoid of any responsibility for the subcontractors actions. Essentially the named contractor became a domestic subcontractor of the principal contractors.

The manner in which named subcontractors are dealt with is largely laid out within the JCT NAM/T and NAM/SC subcontract agreements.

Within the JCT documents the intended subcontractor is named within the tender documents and it is the responsibility of the principal contractor to price the work (or obtain one from the subcontractor), then include all necessary attendances and profits. If or when the main contractor accepts the tender of the named contractor then for all intents and purposes they become a domestic subcontractor.

However, an agreement must still be made between the subcontractor and principal contractor. If there is a failure in this agreement prior to entering into a contract, then the legal responsibility partially falls back onto the client or contract administrator to resolve the problem. It should be noted that such a resolution is only necessary if the problem stems from the particulars of the contract documentation. If the problem arises from the attitudes of one of the contractors then they are liable for the delays.

A similar method is available through the IFC 84 contract. The work to be undertaken by the named contractor is included as a provisional sum. The principal contractor then receives an instruction to accept the offer from a particular contractor, and they have 14 days to reach an agreement.

Nominated

Nominated subcontractors are principally subcontractors selected by the client then transferred over to the main contractor to execute the work. This is often done for works that are critical to the client, either technically or through extended time periods that required organisation prior to the appointment of the principal contractor.

Nomination does provide an effective method of managing the supply chain. The client is able to execute a procurement strategy on a range of work packages free from the involvement of other parties. The client is able to select these parties on their own balanced criteria of price, quality and time, and adjust this weighting for each works package. This can be carried out at any time during the design and construction phase, allowing for the early involvement of critical subcontractors.

This method allows the client to fully select all critical members of the supply chain develop a unique strategy for each works package, determine the level of competition or technical assessment required, procure them on individual criteria of cost, time and quality, and finally bring them on board the project at any time. The selected subcontractor can then be transferred over to the main contractor who then takes on the majority of the legal responsibility for the subcontractors actions on site.

The above appears to be an ideal procurement methodology. However, it is not ideal. Typical problems associated with nomination include the facts that:

- the client is dependent on their own or advisor's knowledge of the marketplace;
- the nominated subcontractor may not be compatible or acceptable to the prime contractor;
- depending upon the specific contractor used, the client or contractor administrator may still have legal responsibilities for the subcontractor's performance;
- the process can be long and complex, involving detailed procedures and interlocking contracts.

These problems, however, must be balanced with the additional benefits that can accrue through the use of specific contracts, namely that:

- the principal contractor receives an assured level of profit and attendances on the work;
- the responsibility of default of the subcontractors is largely removed from the contractor;
- under certain contracts, such as the JCT 80, interim payment certificates govern payment to the nominated subcontractors, resulting in assured payment to them.

It is also possible to have a nominated supplier who merely supplies materials. A prime cost or provisional sum within the contract is normally used to cover their work.

Direct to End Client

Under certain procurement arrangements, namely construction management, the subcontractor can enter into a contract directly with the client. This is obviously an entirely different type of contract relationship, as the contractor is no longer a subcontractor on the same understanding as in the other arrangements. Essentially they become their own primary contractor and therefore must come under the guise of a contract that contains all the necessary provisions to properly arrange and execute such a contract. Under a construction management arrangement there will obviously be some provisions for co-ordination of the other contractors to make the entire process workable.

There are no standard forms of agreement between the various work package contractors and the client. It is usual for the construction manager to supply the client with a suitable contract, normally a heavily modified JCT contract. The revisions are necessary to make the contract workable under a direct contract with the client and provide sufficient authority to the construction manager to properly manage the works. As most clients who use construction management are extremely experienced, they would normally rely on contracts specially developed for their own purposes.

Where contractors are hired by a client in any capacity other than construction management it would be normal practice to use a standard contract such as those produced by the JCT for Minor Works or similar.

3.2.3 Nature of Consultancy Agreements

Contracts for consultancy services differ greatly from those for construction installations. Whereas installation contracts are the embodiment of roles and responsibilities required of the contractor to execute set work defined in accompanying documents, such as drawings and specifications, consultancy agreements act as a combined document to both define the work and set the conditions of the contract.

This difference in nature is perhaps the first cause of problems within the procurement of building services. Whereas the contractors have precisely defined objectives set out within the drawings and specifications, the consultant's role is merely bound by a set of vague duties. The only manner in which this can be overcome is to change the procurement strategy and place both the design and construction with a single body. This is not always desirable and therefore careful drafting of the contracts is required.

Not only is there a conflict between contractor and consultant, but differing terms of engagement between consultants can also be a problem. It is therefore recommended that one set of standard terms and conditions should serve as the framework for all consultancy appointments.

The general methods of procurement and their inherent roles and responsibilities are best reflected in the terms and conditions set out within the Association for Consulting Engineers Conditions of Engagement (1995). The three general methods of procurement are reflected in the document's respective titles:

1. Agreement A(2) – Consulting Engineer Engaged as Lead Consultant
2. Agreement B(2) – Consulting Engineer Engaged Directly by Client, not as Lead Consultant
3. Agreement C(2) – for use where a Consulting Engineer is Engaged to Provide Design Services for a Design and Construct Contractor.

All of the documents use the RIBA (Royal Institute of British Architects) Plan of Work to outline the responsibilities by work stage section:

A Appraisal Stage

Identification of Client's requirements and of possible constraints on development. Preparation of studies to enable the Client to decide whether to proceed and to select the probable procurement method.

B Strategic Briefing

Preparation of Strategic Brief confirming key requirements and constraints. Identification of procedures, organisational structure and range of consultants and others to be engaged for the Project.

C Outline Proposals

Commence development of Strategic Brief into full Project Brief. Preparation of Outline Proposals and estimate of cost. Review of procurement route.

D Detailed Proposals

Complete development of the Project Brief. Preparation of Detailed Proposals. Application for full Development Control approval.

E Final Proposals

Preparation of final proposals for the Project sufficient for co-ordination of all components and elements of the Project.

F Production Information

F1 Preparation of production information in sufficient detail to enable a tender or tenders to be obtained.

F2 Preparation of further production information required under the building contract.

G Tender Documentation

Preparation and collation of tender documentation in sufficient detail to enable a tender or tenders to be obtained for the construction of the Project.

H Tender Action

Identification and evaluation of potential Contractors and/or Specialists for the construction of the Project. Obtaining and appraising tenders and submission of recommendations to the Client.

J Mobilisation

Letting the building contract, appointing the contractor. Issuing of production information to the contractor. Arranging site handover to the Contractor.

K Construction to Practical Completion

Administration of the building contract up to and including practical completion. Provision to the Contractor of further information as and when reasonably required.

L After Practical Completion

Administration of the building contract after practical completion. Making final inspections and settling the final account.

Although the RIBA Plan of Work is heavily criticised for not truly representing the cyclical nature of design, as yet no superior description has been put forward. What is does provide, for both the client and consultant, is a convenient manner of attributing fees to both the quantity of work and its timely delivery.

Criticism has also been levelled against such precise statements of work for limiting both the creative flair of the designer and the necessary teamwork spirit now required in modern business.

The consultant's role, however described, is to translate the overall objectives of the client into a set of instructions that can be executed by the contractor. To deliver the greatest value of service requires not just engineering excellence, but a talent for innovation, flair and teamwork. This requires a very fluid approach in duties and timings to ensure that the overall building, and not just the services, acts in the manner prescribed by the client. Ideally all the project consultants and contractors involved with the design element should have sufficient freedom in their roles to act as a single design team, rather that merely executing a specific element. It is for this very reason procurement must be strategic and embody a number of the modern management philosophies.

Design Responsibilities

The consultant's main reason for appointment under any procurement strategy is to undertake the independent design of the installation. Therefore, when determining the conditions of contract, the clear allocation for design responsibilities is paramount. It is paramount also in building services as the most comprehensive design project will still have areas where it is the contractor's responsibility to complete the design.

This is further complicated by the interrelationship between the services and the building envelope. In the past both could be designed separately - and quite often were -, but modern services demand a dynamic relationship with the rest of the building. Therefore, the allocation of design responsibilities has become even more complex.

It is for this very reason that it is imperative that a comprehensive schedule of design responsibilities is agreed between all parties of the building services team and the project team. Detailed guidance for doing so is included in Chapter 13 of this text.

3.2.4 Relationship with Procurement

There is a lengthy and needless debate as to whether the procurement arrangement is determined by the contract. This debate is often put forward by ill-informed people who do not fully understand the nature of procurement and its relationship with the contract.

A contract will largely predetermine:

- the clarity with which each participant's tasks are designed
- the adequacy of information and the ease with which it is accessed
- the interdependence of activities with other parties.

A procurement arrangement is determined by the overall strategy developed to fulfil a client's requirements through design, the acquisition of resources and the execution of the design into the intended structure. The contract allows the fine tuning of this strategy to provide the necessary tactics to manipulate the parties. The contract should never be relied upon to correct a poorly developed strategy or as the only means to control the project. As to whether the contract is adversarial or not is determined by the political environment of the project.

To procure building services properly requires the following to be clearly developed within a strategy prior to the drafting of any contracts:

- a clear statement of responsibilities as to who will design what;
- a comprehensive schedule of information required by all parties, stating when and in what format it is to be supplied;
- a properly developed design, procurement and construction programme;
- a decisive brief that clearly states the technology involved, its reasons for use and performance benchmarks;

- a commissioning and handover strategy;
- a quality plan for design, installation, testing and commissioning/handover;
- a detailed cost plan.

Although progress is being made with the contracts between main contractor and client to embody the necessary provisions and attitudes to achieve a successful project, little progress has been made with specialist engineering contractors.

Several key issues are at the heart of the debate as to whether contracts are fair. These include clauses for prompt payment, clear identification of design responsibilities, secured payment from the end client to the specialist contractor, quick and fair dispute resolution and either the abolition or modification of the traditional retention procedures.

REFERENCES AND FURTHER READING

1. Association for Consulting Engineers, (1995) *Association for Consulting Engineers Conditions of Engagement*, London, ACE Publications.
2. Ashworth, A. (2001) *Contractual Procedures*, London, Pearson Education.
3. Chappell, D. and Powell-Smith, V. (1994) *Building Sub-Contract Documentation*, London, Blackwell Scientific Publications.
4. HSE, (1974) *Health and Safety at Work Act*, London, HMSO.
5. Latham, M. (1984) *Constructing The Team*, London, HMSO.
6. Murdoch, J. and Hughes, W. (2000) *Construction Contracts Law and Management*, London, E&FN Spon.
7. RIBA (1992) *The Architects Job Handbook*, London, RIBA Publications.

Chapter Four

Project Modalities

4.1 PROJECT MODALITIES

The first set of modalities, as outlined in Figure 4.1, present the initial stages of the project. These are normally concerned with establishing the need for the project and financial affordability of it.

Building services will rarely influence the need, other than extreme redundancy of the existing system makes the affordability of the existing building untenable. These are normally global issues based on the business strategy for the company involved.

The first major influence for building services comes in the business case of the options assessment, as it plays a major part in both construction costs and operating cost of the new building.

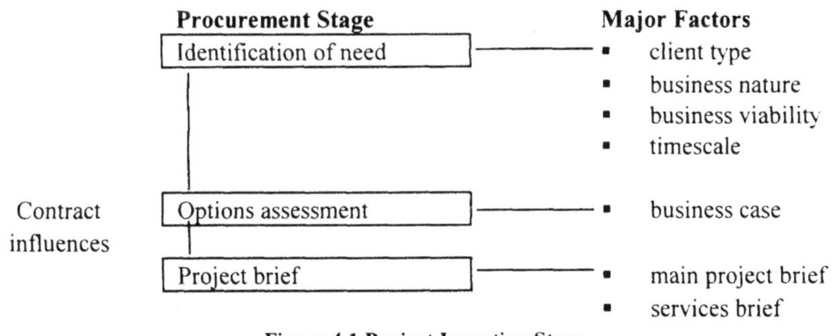

Figure 4.1 Project Inception Stage

The modalities presented here concentrate on the project brief. They provide sufficient checklists to ensure the brief adequately covers the client's needs. The other modality covered is the appropriate selection of the contract. Although within the text that the concentration at procurement stage on contracts is unfounded, its overall influence on how the project is to be delivered must not be ignored, especially when reviewing the influence of risk on the overall project.

4.2 THE BUSINESS CASE

Any construction project is an investment that must justify its existence. This is normally done through a business case. These calculations are often complex and longwinded, taking into account such diverse cost categories as land cost, tax rebates and taxation, balanced against the expected return of both rental income and capital growth. The construction cost will often account for only 10-25% of the total project cost.

The building services proportion of the calculation is obviously very small, but is becoming increasingly more important. Moving from the macro-sized project business case to the micro case of the actual building, building services is often the dominate element. The detail in which services are addressed separately will depend upon the client motivation. For example, speculative developers will only consider capital costs, while owner-occupiers will balance capital cost with operating cost.

The business case for building services must consider three aspects:

- **Capital Cost:** the initial investment required to construct or refurbish the installation. This will include design fees, contractor costs, incoming service supply installations (gas, water, electrics etc.). To encourage the investment in facilities, the government provides for capital tax allowances which permit the cost of certain items of plant to be deducted from the business taxes payable by the client.
- **Operating Cost:** will include the fuel cost to power the system, maintenance cost and associated cost such as attendant labour to properly operate the system. With the increasing focus on sustainability, the UK government is increasing the tax on fuels through the Carbon Tax and the maintenance of buildings through building MOT's and certified commissioning.
- **Quality/Cost Comparison:** no client can afford the ideal system. The business case must assess each possible option and put forward the most suitable, based upon the available budget. This can be carried by one of two ways: residual valuation subtracts all costs from the available budget showing the remainder as the amount available that can be afforded for the design. The return basis considers all costs as an investment and will compare the return against other investments, such as stocks or bonds.

Building services costs should not be considered in isolation. A building is a compete entity and where one element is changed, it will influence the others. For example when considering air conditioning, it will influence the following:

- increase in electrical services
- double the needed plant room size

- increase in ceiling void area for ductwork, causing an increase in storey height
- influence fenestration design to reduce solar gain
- influence envelope design to minimise air infiltration.

Figure 4.2 provides the summary for cost categories. Greater detail of each cost can be found in Chapter 8 under whole life costs.

Business Case Pro-Forma				
Project :				
Option assessment :				
Assumptions :				
Cost Category	**Headline Costs**	**Assumption**	**Capital Cost**	**Whole life Cost**
Site acquisition	services diversion			
	incoming services requirements			
Demolition	existing services			
	remedial works			
Construction costs	installation costs			
	design fees			
	bwic services			
Operating costs (inc fuel)	heating			
	cooling			
	public health			
	power			
	lighting			
Maintenance	heating			
	cooling			
	public health			
	power			
	lighting			
Fees	design			
	consultancy			
	commissioning			
	commercial			
Taxes	capital allowances			
	carbon tax			
	VAT			
Administration	temp. services			
	duplex services			
	relocation			
	upgrades			
		Total		

Figure 4.2 Pro-forma Business Case Summary

4.3 BRIEFING

4.3.1 Design Parameters

The client and services design team should consider the following statements in Table 4.1 with due regard to current and future occupancy levels, together with any anticipated changes in legislation or social patterns.

Building Occupancy Details

- maximum number of occupants/visitors
- distribution of occupants
- proportions of male to female occupants
- occupancy archetype – hours and use patterns
- out-of-hours use.

Building Plant Loads

- fuel preference
- load demands
- gas burning equipment including kW loading
- electrical equipment including kW loading
- usage patterns of plant and equipment
- equipment needing supply from building services – e.g. medical gases
- extract or ventilation requirement, including industrial process machinery
- noise levels of plant
- trade effluent and treatment requirements
- specialist security arrangements.

Environmental Performance Criteria

- room by room list of acceptable values for:
  - dry bulb temperature
  - humidity
  - lighting level
  - noise rating
  - minimum fresh air requirement
- window shading details.

Building Structure

- U-values of building envelope
- flexibility requirements of room partitions
- anticipated levels of office reorganisation
- flexibility requirements of desk positions.

Environmental Control

- areas of building requiring close control
- areas of building needing direct control by occupants
- areas of building with plants/animals needing particular environmental control.

Domestic Water Supplies

- requirement for main water other than for drinking water purposes
- chilled drinking service requirements
- watering point requirements
- vehicle washing points
- drink vending machine requirements.

Electrical Services

- client preference in lighting fitting selection
- equipment needing three phase supply
- workplace unit power requirements
- telephone power requirements
- external/landscaping lighting and power requirements
- flameproof electrical services
- power and service needs for:
  - public address system
  - special socket outlets
  - illuminated sign requirements
  - overhead doors
  - fire / security alarms
  - lifts.

Data Cabling

- cabling requirements
- trunking
- cable management
- floor box requirements.

Stand-by Power

- equipment needing stand-by power
- stand-by power location
- fuel requirements.

Special Areas

- kitchens
- conference facilities
- computer rooms
- clean rooms
- industrial areas
- underground garages.

Maintenance

- preference for low maintenance equipment
- disruption tolerance for maintenance works
- maintenance requirements and responsibilities during defects period
- plant and equipment labelling.

Table 4.1 Functional Requirements

4.3.2 System Selection

The majority of systems that come within the general category of building services are listed in Table 4.2. The list can be used as a 'menu' from which to select the systems appropriate to the building under consideration. Some system headings have additional request for information, which the designer will need in order to commence the design for those systems.

Disposal Systems

Drainage

- rainwater pipework/gutters
- foul drainage above ground
- drainage below ground
- land drainage
- laboratory/industrial waste drainage

Sewerage

- sewage pumping
- sewage treatment/sterilisation

Refuse disposal

- centralised vacuum cleaning
- refuse chutes
- compactors/macerators
- incineration plant

Piped Supply System

Water supply

- cold water
- hot water
- hot and cold water (small scale)
- pressurised water
- irrigation
- fountains/water features

Treated water supply

- treated/deionised/distilled
- swimming pool water treatment

Gas supply

- compressed air
- instrument air
- natural gas
- liquid petroleum gas
- medical/laboratory gas

Petrol/Oil storage

- petrol/oil – lubrication
- fuel oil storage/distribution

Other supply systems

- vacuum
- steam

Fire fighting – water

- fire hose reels
- dry risers
- wet risers
- sprinklers
- deluge
- fire hydrants

Fire fighting – gas/foam

- gas fire fighting
- foam fire fighting

Mechanical Heating/Cooling/Refrigeration Systems

Heat source

- gas/oil fired boilers
- coal fired boilers
- electrode/direct electric boilers
- packaged steam generators
- heat pumps
- solar collectors
- alternative fuel boilers

Primary heat distribution

- heating
- cooling

Heat distribution/utilisation – air

- medium temperature hot water heating/low temperature hot water heating/steam heating

Heat distribution/utilisation – air

- warm air heating
- warm air heating (small scale) local heating units

Heat Recovery

Central refrigeration/distribution

- central refrigeration plant
- primary/secondary cooling distribution

Local cooling/refrigeration

- local cooling units
- cold rooms
- ice pads

Ventilation/Air Conditioning Systems

Ventilation/fume extract

- general supply/extract
- toilet extract
- kitchen extract
- car parking extract
- fume extract
- anaesthetic gas extract
- smoke extract/smoke control
- safety cabinet/fume cupboard extract

Industrial extract

- dust collection

Air conditioning – all air

- low-velocity air conditioning
- VAV air conditioning
- dual-duct air conditioning
- multi-zone air conditioning

Air conditioning – air/water

- induction air conditioning
- fan-coil air conditioning
- terminal re-heat air conditioning
- terminal heat pump air conditioning

Air conditioning – hybrid

- hybrid system air conditioning

Air conditioning – local

- free standing air conditioning units
- window/wall air conditioning units

Other air systems

- air curtains

Electrical Supply/Power/Lighting Systems

Generation/Supply/HV distribution

- electricity generations plant
- HV supply/distribution/public utility supply
- LV supply/public utility supply

General LV Distribution/Lighting/Power

- LV distribution
- general LV power

- general lighting

Supply types

- extra low voltage supply
- DC supply
- uninterrupted power supply

Special lighting

- emergency lighting
- street/area/floodlighting
- studio/auditorium/area lighting

Electric heating

- electric underfloor heating
- local electric heating units

General/Other electrical work

- general lighting and power (small scale)

Communications/Security/Control Systems

Communications – speech/audio

- telecommunications
- staff paging/location
- public address/sound amplification
- centralised dictation

Communications – audio-visual

- radio/TV/CCTV
- projection
- advertising display
- clocks

Communications – data

- data transmission

Security

- access control
- security detection and alarm

Protection

- fire detection and alarm
- earthing and bonding
- lightning protection
- electromagnetic screening

Control

- monitoring
- central control
- building automation

Transport Systems

People/Goods

- lifts
- escalators
- moving pavements
- goods hoists

Goods/Maintenance

- hoists
- cranes
- travelling cradles
- goods distribution/mechanised warehousing

Documents

- mechanical document conveying
- pneumatic document conveying
- automatic filing and retrieval

Table 4.2 System Selection

4.4 CONTRACT SELECTION

Selecting the appropriate contract is paramount in ensuring a successful project. When developing an appropriate strategy for building services procurement, it is important to understand how this selection will limit the choices available and affect the relationships between the project parties.

4.4.1 Determining Priorities of Project Objectives

All contracts influence the primary objectives of cost, time and quality differently. The first stage in determining a suitable contract is to understand the possible project priority profile and how this matches possible contracts (see Figure 4.3).

Project Priority Profile		priorities: 1=low, 5=high				
Cost Priorities						
1	cost certainty prior to project commence	1	2	3	4	5
2	cost certainty as a lump sum for work	1	2	3	4	5
3	requirement for cost accountability	1	2	3	4	5
4	award criteria - price alone	1	2	3	4	5
5	award criteria - quality alone	1	2	3	4	5
6	award criteria - price/quality mechanism	1	2	3	4	5
7	need to vary work amounts while maintaining cost accountability	1	2	3	4	5
8	price fluctuations - permissible	1	2	3	4	5
9	cost balance - cost only	1	2	3	4	5
10	cost balance - whole life and capital cost	1	2	3	4	5
Time Priorities						
1	time requirement - greater than quality	1	2	3	4	5
2	time requirement - greater than cost	1	2	3	4	5
3	phased completion required	1	2	3	4	5
4	criticality of commencement date	1	2	3	4	5
5	criticality of end date	1	2	3	4	5
Quality Priorities						
1	quality requirements - greater than cost	1	2	3	4	5
2	quality requirements - greater than time	1	2	3	4	5
3	is quality esoteric	1	2	3	4	5
4	is quality highly specialised	1	2	3	4	5
5	building services quality compared to rest of project	1	2	3	4	5
6	need for direct control over quality	1	2	3	4	5
7	need for independent assessment of quality	1	2	3	4	5

Note: there is no standard contract that takes into account whole life cost

Figure 4.3 Project Priority Profile

The results of this selection should then be compared against the priority profile of standard contracts shown in Figure 4.4.

Contract Selection Chart			
Contract Choice	**Cost**	**Time**	**Quality**
JCT Contracts			
Standard Form With Quantities	▪ high accountability ▪ high certainty	▪ slow due to preparation of quantities	▪ high only if correctly specified
Standard Form Without Quantities	▪ low if work is likely to be varied	▪ medium	▪ high only if correctly specified
Intermediate Form of Building Contract	▪ dependent on pricing method and as to whether specifications or bill of quantities are used		
Agreement for Minor Building Works	▪ high as work must be simple and set against lump sum	▪ quick as form is simple	▪ not suitable for complex work
Standard From With Approximate Quantities	▪ low as quantities are approximate, but provides accountability	▪ medium as quantities are approximate	▪ dependent on specification and management of the works
Standard Form of Management Contract	▪ low as normally based on fees on prime cost	▪ high as work stages over-lapped	▪ high as supply chain can be involved
Fixed Fee Form of Prime Cost Contract	▪ medium as fee is fixed, but work based on prime cost	▪ high as work stages over-lapped	▪ high as supply chain can be involved
Standard Form With Contractor's Design	▪ low if work is likely to be varied	▪ medium	▪ medium, depending upon design
Standard Form with Quantities Modified by Contractor's Designed Portion Supplement	▪ medium as quantities allow accountability, but difficult on designed elements	▪ medium	▪ medium, depending upon design
Other Contracts			
Engineering and Construction Contract	▪ high as detailed costs can be submitted in a variety of formats	▪ high as work stages can be overlapped	▪ high as supply chain can be involved
GC/Works 1	▪ medium to high depending upon pricing option	▪ medium depending upon option	▪ medium depending upon option

Figure 4.4 Contract Selection Chart

Except in the case of a management type of contract all building services contractors are contracted to the main contractor through a subcontract arrangement. Standard forms of contract exist that are specially designed for use with the above contracts. Figures 4.5 and 4.6 summarise the main choices and differences between these contracts.

Major Forms of Subcontract	
Form of Subcontract	**Use**
Domestic Form of Subcontract DOM/1	For use when the subcontract is purely a domestic arrangement under a JCT contract
Domestic Form of Subcontract DOM/2	For use when a domestic subcontractor accepts the responsibility for an element of design under a JCT contract
Intermediate Subcontract NAM/SC	For use when the subcontractor is named within a JCT main contract
Nominated Subcontract NSC/C	For use when the contractor is nominated under a JCT contract

Figure 4.5 Main Forms of Subcontract

Contract Difference			
Major Issue	**Contract/Clause**		
	DOM/1/2	**NSC/C**	**NAM/SC**
Right to arbitration	yes	yes	yes
Notice of discrepancies	must report	must report	not required to report
Right to refuse instruction	no right	no right	able to refuse
Right for main contractor to refuse named / nominated party	n/a	m/c has right to refuse	m/c must accept at tender stage or refuse if based on instruction
Ability to work before signing of contract	no provision	employer has right to preorder	no provision
Bonds and warranties to be provided	yes	yes	yes
Design liability direct to employer	no	yes	yes if CASEC form signed
Right to instruct main contractor to use particular company	no	yes	yes
Compliance with programme	reasonably progress the work	time inserted in contract	time inserted in contract
Independent valuing of work	n/a	by quantity surveyor	n/a
Failure of payment	right to suspend work	right to be paid direct by employer	right to suspend work

Figure 4.6 Contract Differences

4.5 RISK MANAGEMENT

All projects will carry some form of construction risk. In the early stages of the development of a procurement strategy it is necessary to understand the general sources of risk and how each possible option for design and procurement will affect them.

The manner and extent to which risks are evaluated will vary through the project. At the early stages the focus must remain at an overview level. What is trying to be understood here is the global issues that will affect the project. It is not necessary to undertake the detailed quantification of risk consequences at this time as it is likely that sufficient information to do available this will not be available.

Figure 4.7 provides a guide to the type of risks likely to be encountered upon a normal project. Figures 4.8 and 4.9 provide two methods of determining the risk level of the project. While Figure 4.10 provides the general method in which the responses should be determined.

Stage	**Risk Potential, Source and Level**		
	Low	**Medium**	**High**
Post Handover		technology inappropriate to function	→
		inadequate maintenance	→
		latent defects of manufacture/installation	→
		unstable performance	→
		systems operating too near to limits	→
BMS and Controls		software unproven/fails to work properly	→
		system fails to work properly	→
	possible manufacturing/installation defects		→
	exposed		→
Regulation		poor system preparation resulting in	→
Systems Preparation		poor system layout or poor installation	→
	installation defects		→
	component design defects		→
Installation		interface problems between service systems and building structure/fabric	→
	learning curve for new technology installation		→
Prestart		use of incorrect procedures	→
	discovery of risks during design		→
	failure to achieve design programme		→
Pre-award		latent risk due to concentration on commercial issues	→
	selection of appropriate contractors/designers		→
	market conditions change from anticipated		→
	location of proper component suppliers		→

Figure 4.7 Potential Sources of Risk

Project Risk Assessment

Overall Risk Assessment

☐ Normal

☐ High

Risk Consideration	Criteria	Risk Assessment low	med	high	Proposed Response
Project Environment					
User Organisation	Stable/Competent	❑	❑	❑	
	Inexperienced	❑	❑	❑	
User Management	Works as a Team	❑	❑	❑	
	Factions/Conflicts	❑	❑	❑	
Joint Venture	Clients Sole Contractor	❑	❑	❑	
	Third Party Involved	❑	❑	❑	
Public Visibility	Little/None	❑	❑	❑	
	Significant/Sensitive	❑	❑	❑	
No. of Project Sites	2 or less	❑	❑	❑	
	3 or more	❑	❑	❑	
Environmental Impact	High	❑	❑	❑	
	Low	❑	❑	❑	
Project Management					
Involvement	Active Involvement	❑	❑	❑	
	Limited Participation	❑	❑	❑	
Experience	Strong Experience	❑	❑	❑	
	Weak Experience	❑	❑	❑	
Participation	Active Participation	❑	❑	❑	
	Limited Participation	❑	❑	❑	
Project Manager	Experienced/Full Time	❑	❑	❑	
	Unqualified/Part Time	❑	❑	❑	
Techniques	Effective Techniques	❑	❑	❑	
	Ineffective Techniques	❑	❑	❑	
Clients Experience	Experienced Client	❑	❑	❑	
	Inexperienced Client	❑	❑	❑	
Project Characteristics					
Complexity	Straightforward	❑	❑	❑	
	Complex	❑	❑	❑	
Technology	Proved/Traditional	❑	❑	❑	
	New Tech./Complex	❑	❑	❑	
Impact of Failure	Minimal	❑	❑	❑	
	Significant	❑	❑	❑	
Scope	Typical	❑	❑	❑	
	Unusual	❑	❑	❑	
Type	New Construction	❑	❑	❑	

	Refurbishment	❑	❑	❑
User Acceptance	Strong Support	❑	❑	❑
	Controversial	❑	❑	❑
Proposed Time	Reasonable Allowance	❑	❑	❑
	Rapid	❑	❑	❑
Schedule Completion	Flexible	❑	❑	❑
	Absolute	❑	❑	❑
Potential Changes	Stable	❑	❑	❑
	Dynamic	❑	❑	❑
Man-days	less than 1000	❑	❑	❑
	1000 or more	❑	❑	❑
Cost-benefit Analysis	Proven Methods	❑	❑	❑
	Inappropriate	❑	❑	❑
Project Staffing				
User Participation	Active	❑	❑	❑
	Passive	❑	❑	❑
Project Supervision	Meets Standards	❑	❑	❑
	Below Standards	❑	❑	❑
Project Team	Adequate Skills/Exper.	❑	❑	❑
	Low Skills/Exper.	❑	❑	❑
Project Costs				
Cost Quotation	Fixed Price	❑	❑	❑
	Re-measurable	❑	❑	❑
	Service Agreement	❑	❑	❑
Cost Estimate	Bill of Quants.	❑	❑	❑
	Spec. and Drawings	❑	❑	❑
	Design and Engineer	❑	❑	❑
Form of Contract	Standard	❑	❑	❑
	Modified Standard	❑	❑	❑
	Bespoke	❑	❑	❑
Other				
As Applicable		❑	❑	❑

Figure 4.8 Project Risk Assessment

Figure 4.8 provides a simple manner in which to categorise the risk level of a project. It not only gives the overall project risk rating but shows the areas in which high level of risks are likely to occur.

The assessment though is highly subjective as it requires considerable expertise to assess what is high risk and whether it exists within the project. Ideally this chart should be completed by the building services team within a facilitated workshop. One method of completing it is to analyse a previous project as to the level of risk actually incurred on the project during its execution. This can then be used as a benchmark for the project being completed.

Figure 4.9 provides for a more quantitative approach to assessing risk as the risk levels are largely predetermined. Projects scoring in excess of 16 should be considered as high risk.

Project Risk Assessment		**Overall Project Value £m**			
Heading	**Consideration**	**<2.5**	**2.5 -10**	**10 –20**	**>20**
Proportional Value of Services	<15%	0	2	4	6
	15-35%	2	4	0	0
	>35%	6	8	0	0
Type of Building	Offices	1	1	1	1
	Factory	0	0	1	1
	Retail	0	0	1	2
	Residential	0	0	1	2
	Hospital	1	1	2	4
	Other	1	1	2	4
Air Conditioning Included	Yes	6	8	4	2
	No	0	0	0	0
Type of Air Conditioning	VAV	0	0	0	0
	VRV VRF	1	1	0	0
	Chilled Beams	2	2	1	1
	Displacement Vent	2	2	1	1
Temp. Performance Criteria	$\pm 2^0$C	0	0	0	0
	$\pm 1^0$C	1	1	1	1
	$\pm 0.5^0$C	2	2	2	2
Humidity Performance Criteria	± 20%RH	0	0	0	0
	± 10%RH	2	0	0	0
	± 5%RH	3	3	3	3
Noise Performance Criteria	45NR	0	0	0	0
	40NR	1	1	0	0
	35NR	2	2	2	2
	>30NR	4	4	4	4
Complex / Unusual Services	No	0	0	0	0
	Yes	1	3	2	1
Design	Standard	0	0	0	0
	Innovative	1	3	2	1
Scheme Design	One-off	0	0	0	1
	Multiple	0	0	1	4
Electronic Inverters Used	No	0	0	0	0
	Yes	1	1	2	2
BMS or BEMS	No	0	0	0	0
	Yes	1	2	1	1
Plant Space Allowance	Adequate	0	0	0	0
	Confined	1	2	1	1
Air Tightness Required	Room	1	2	1	0
	Building	1	1	2	3

Fume Exhaust Required	No	0	0	0	0
	Yes	1	2	1	0
Designer Capabilities	Suspect	1	1	2	4
	Proficient	1	1	0	0
Construction Programme	Normal	0	0	0	0
	Fast	0	1	0	0
	Very Fast	1	2	1	1
Designer Appointment	Standard	0	0	0	0
	Spec. & Drawings	1	2	0	0
Tender Information	Average	0	0	0	1
	Poor	1	1	2	3
Design Development Required	Extensive	1	2	1	1
	Normal	0	1	0	0
Tenders from Suppliers	Normal	0	0	0	0
	Qualified	0	2	1	0
RISK ASSESSMENT	**TOTAL SCORE**				

Figure 4.9 Project Risk Assessment

Once the risks of project have been identified and the general risk level of the project has been determined, the procurement strategy must reflect these and determine a manner in which to respond best.

The contract provides the vehicle in which risks are apportioned to parties. The allocating of risks must be done fairly and objectively. The party being allocated the risk must be willing to accept it and be capable of managing it.

Figure 4.10 shows the general manner in which risks should be apportioned. Each risk needs to be analysed for the correct response based on a hierarchy. Four general responses exist: eliminate, minimise, transfer and accept. The process must commence with the most ideal response, which is to eliminate the risk, with the process culminating in the least desirable response, which is to accept the risk. Although at this point the risks should have been minimised as much a possible.

Once a response is determined a cyclical process begins where the response must be examined for secondary risks. For example, the transferring of risk to a contractor or designer may appear to eliminate the risk, but the secondary risk is whether that party is capable of handling the risk created. Mismanagement or business failure (caused by the financial burden of the risk) may result in a far larger risk eventuality having to be dealt with.

On a similar basis the nature of risks will change during the life cycle of the project and therefore an effective monitoring process which intermittently carries out the risk-response process again must be enacted.

Building services require a complex risk response as each independent source of risk must be analysed. Generally a risk-response process should be carried out on the following:

Construction risks: are based on the project and stem from the nature of the project. These can include: completing the project on time; completing within budget; health and safety risks associated with the installation; and achieving satisfactory quality levels.

Contractor/Designer risks: are associated with either party being able to effectively execute their roles. These may include their own business solvency and effective operation to the actual competency in which they execute the work. An effective procurement strategy and selection process should help to minimise these risks.

Technology risks: building services by their very nature are complex. Each project is a prototype and therefore the risk of successful operation is inherent. Not only must the technology be reviewed for expected success levels, but the effective level of health and safety associated with the operation needs to be reviewed.

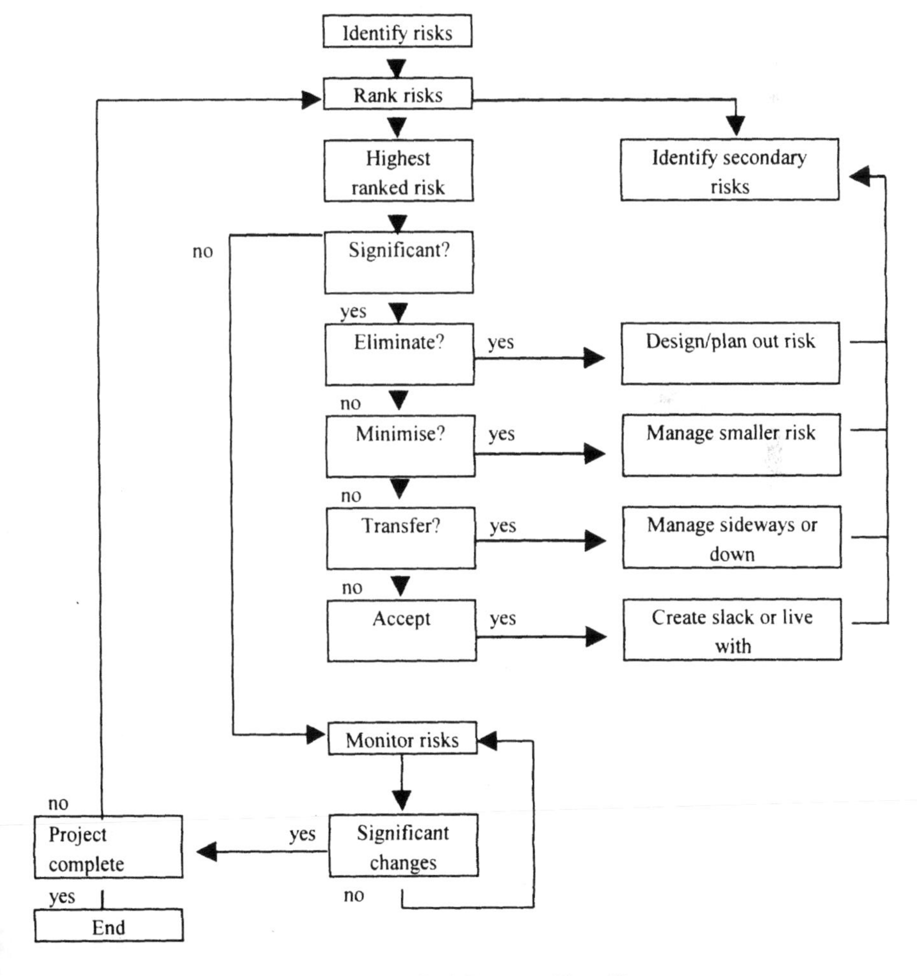

Figure 4.10 Risk-Response Flow Chart

REFERENCES

1. Ashworth, A. (2001) *Contractual Procedures*, London, Pearson Education.
2. Chapman, C.B., Ward, S.C. and McDonald, M. (1989) *Roles, Responsibilities and Risks in Management Contracts*, SERC Research Grant Report, University of Southampton.
3. Chappell, D. and Powell-Smith, V. (1994) *Building Sub-Contract Documentation*, London, Blackwell Scientific Publications.
4. CIOB (1996) *Code of Practice for Project Management for Construction and Development,* Ascot, Chartered Institute of Building.
5. Cox, S. and Clamp, H. (1999) *Which Contract?,* London, RIBA Publications.
6. Murdoch, J. and Hughes, W. (2000) *Construction Contracts Law and Management*, London, E&FN Spon.
7. RIBA (1992) *The Architects Job Handbook*, London, RIBA Publications.
8. Wild, J. (1997) *Site Management of Building Services Contractors,* London, E&FN Spon Limited.

Chapter Five

Developing a Strategy

5.1 THE NATURE OF STRATEGY

Prior to considering the detailed nature of procurement, a product or scheme must be developed. This is the amalgam of the client's ideas as to how the building services should operate in terms of technical performance and cost. As installations become more complex, together with a growing sophistication in the market place as to how to best deliver a client's needs, other issues have begun to enter the arena of considerations. Modern ideas such as brand strategies, whole life performance and service level agreements have changed the very nature of a building services installation.

Much has been written as to the principles of strategy, the environment that shapes it and the nature of decision making. It is not necessary to elaborate on the philosophical debates that are used in countless management texts as to how a strategy should be developed and executed. Within the context of procurement it is sufficient to understand that strategy consists of management principles that are enabled to achieve a given set of objectives. These objectives are derived from the client's overall reason or mission for undertaking a building services installation. The strategy enables a set of actions to be created that serve to empower the building services team. These concepts are best thought of as a hierarchical pyramid.

Mission	: sets the end goal
Objectives	: facts that support the mission
Goals	: deliverables of the mission
Strategy	: how the mission is to be achieved
Actions	: specific events undertaken

Figure 5.1 Strategy Hierarchy

Diagramatically it can be seen that in order to achieve a strategy the overall end objectives of the project must first be determined. For building services, this is normally expressed as the key benchmarks of time, cost and quality, with secondary benchmarks being technical performance, maintenance, operability, life cycle and adaptability.

These end goals can be subdivided into the three main headings of design, cost and product strategies.

5.2 DESIGN STRATEGY

The purpose of design is to construct a workable arrangement of technology that will deliver the technical objectives of the project. Unlike architecture, building services must deliver the described technical performance exactly. To under-deliver leaves the client with a building that is unusable, to over-deliver leaves the client with a building that is too expensive or complex to operate.

To use the example of a refrigerated warehouse, if the clients end objective is to achieve -3.0^{0}C, then -2.5^{0}C is technically unacceptable. Building services performance is largely non-arbitrary, and therefore the primary consideration in design is what technical strategy is available that will deliver the client's end objective. This developed strategy will be supported by the quality level required, which itself is driven by the life cycle of the proposed installation. Cost is merely a consequence of these other decisions. To reiterate, building service's primary function is to achieve prescribed technical objectives.

A more esoteric consideration arises on entering the arena of design, that of brand strategies. A client can manipulate the value of their building by attaching brand labels to both the design and technical components. Much like the line "its got Intel inside" for computers, the same strategy is entering building services.

Therefore the engineer must now straddle the differing analytical requirements of technical excellence and the subjective philosophies of brands and value.

5.2.1 Technology Considerations

Buildings have become a holistic entity, where the envelope, internal environment and the surrounding environment have become a complex and intertwined set of considerations. Architects and building services designers must now work together to develop a composite environment combining aesthetic satisfaction with optimum thermal comfort. There is a greater awareness that the traditional conflicting requirements of aesthetic design with technical performance must be considered in unison. Previously technology was used like a blunderbuss to rectify a design that conflicted with the laws of natural environment. Building orientation, window placement and exploitation of the basic laws of nature have provided better designed and less complex buildings.

As a paradox the overall technology involved with a building services installation has multiplied, both in complexity and number of individual systems. The proliferation of air conditioning and information and control technologies now requires careful consideration within the business case and design brief.

A lack of strategic development for services technology often compromises the effective design of services. Inadequate plant rooms and poorly sited fresh air intakes, combined with poor architectural design for example often results in complex and expensive services design. The client ends up paying severe penalties in both technical performance and operating costs. This underlines the problems of trying to develop a procurement strategy where the overall strategy must achieve a set of end objectives that have become complicated through poorly co-ordinated or inappropriate designs.

A few simple changes to the design parameters may allow use of standard, rather than special, components without compromising design, and often provides higher efficiencies and both capital and operating cost benefits. Under such circumstances the procurement strategy must allow specialist contractors and component suppliers, who hold detailed knowledge of these areas, to participate in the design, while the project design team must be similarly encouraged to react positively to accept their advice.

Service Life

A key element in developing the necessary technology strategy for an installation is to determine the time horizon in which the services must exist. Time horizons vary considerably, and are dependent on both the business case and the clients strategy for the building. Establishing the correct time horizon is critical for both costing and setting the appropriate specification level.

The life of a building or services installation is a complex expression involving many considerations. Ideally the lives of various elements of the building should be predicted from observed data on failure, but this type of information is rarely available. Building life can be extended by periodic maintenance and replacement or may be foreshortened by changing economic, social or legal conditions. There must be a relationship between the expected life of the building, and those of the services.

Obsolescence will determine when the installation has reached the end of its serviceable life. Forms of obsolescence include the following:

- **Physical life** of the building is the period from construction to the time when physical collapse is possible. Although most buildings never reach this point, as they are demolished owing to economic obsolescence, building services can reach this point in a very short time, especially when operating in aggressive environments.
- **Economic life** is the period from construction to economic obsolescence, the period of time over which occupation of a particular building is considered to be the least-cost alternative for meeting a particular objective.
- **Functional life** is the period from construction to the time when the building ceases to function for the same purpose as that for which it was built. Functional and economic obsolescence are therefore closely

interlinked. Functional obsolescence does not normally lead to demolition as the building will often be refurbished to fulfil another function.

- **Technological obsolescence** occurs when the building or component is no longer technologically superior to alternatives and replacement is undertaken because of lower operating costs or greater efficiency.
- **Social and legal obsolescence** occur when human desires dictate replacement for non-economic reasons - where reliability is of the utmost concern, as for example in hospital buildings.

It is difficult to give guidelines as to the "correct" choice of life cycle. There is a strong case in favour of choosing economic or functional life, as the business case of the installation will normally dictate the boundaries of these time periods. When considering the appropriate time line the following points should be noted:

- appropriate economic life will vary with the type of client;
- owner-occupiers might be expected to operate a building in a particular way for rather longer than a rent payer;
- when considering the investment for the public sector client a relatively long time horizon, usually commensurate with the physical life, should be used;
- developers normally seek short term financial gain. In these circumstances the time horizon will equal the client's economic life span for the building, which is the holding period that is expected to maximise speculative profits;
- the discounting process in whole life costing is such that the cost implications of a choice between a 25-year or 30-year life cycle will not, in general, be particularly severe;
- any life calculated purely for taxation and depreciation purposes should not be used for life cycle cost calculations.

Attention must be drawn to the difference between the expected life of a building and the expected lives of the various building components or services installation. Most buildings are structurally capable of lasting for over 100 years. No services installation however can feasibly last for more than 25 years because of technological or economical obsolescence. The life cycle for a building services installation is not the period within which the services will need to be replaced, but rather the operating period of the building for which the services are needed, as they are likely to be replaced several times during the life of the building, and this must be allowed for. On a similar basis individual components will have their own feasible time horizons for replacement, which are often dictated by either maintenance regimes or the desire for interior design changes.

Trends in Technology

Technology is a restless martyr that marches on indefinitely, although in terms of building services technology its progress is rarely considered to be expedient. Nevertheless, what is considered to be up-to-date technology today will always be outdated in the future. Therefore careful consideration must be given as to what technology is appropriate both today and in the future. The more complex the technology the greater the likelihood that it will become dated: there is a general rule of thumb in building services that states that the more complex the technology the quicker it will be superseded by another 'new' technology.

There are unlikely to be major changes in building services technology before the end of the decade. The current drivers of change are mainly sustainability - the seeking of lower energy consumption - and the combining of controls and information technology. There are, however, likely to be greater changes in service delivery. While not a pure technical consideration, the design team must always consider how differing technologies will be delivered to the marketplace.

The issue of sustainability is complex, as it embodies a number of different issues. One of the most dominant is the contribution of refrigerant gases to ozone depletion. The banning of CFCs in the early 1990s is likely to be further hampered by the banning of their replacement HFCs. This likelihood is an example of where service life and technology strategy must be considered simultaneously. Any refrigerant installation that is likely to have a service life of greater than five years must be designed with a consideration of possible gas replacement. Alternatively, a design could be developed that uses different sources of cooling, such as ground water heat pumps, thereby eliminating the risk of gas replacement.

The technology strategy must be made independently of the building services team. Each party within the services team will have a bias towards one technology or another. All will seek the easiest solution to the problem, be it through design or ease of installation. In fact the possible solutions put forward to meet the client's objectives will be limited by the team's knowledge of such systems. Therefore the technology for a project must be developed prior to the team's appointment, thereby ensuring that the technology is the best suited to meet the client's end objectives, and that the team finally selected is the best combination to design and install the system.

An example of this conflict of interests is the commercial success of DX-refrigerant cooling which is often cited as being the result of the ease with which it can be specified. Variable refrigerant volume (VRF) systems, in particular, are seen as a straightforward answer to a cooling problem. The competition of design fees causes the consultant to seek the easiest possible solution. VRF systems are simple and easy to design, with the majority of the work being undertaken by both the installation contractor and component suppliers. These same component suppliers understand this interaction and aggressively market this solution to consultants. However, the result for the client is a compromised design, driven by ease rather than optimality.

5.2.2 Branded Products

Maximising value is a key aspect of any procurement strategy. As value is the resulting relationship between cost and quality, to deliver improved value-for-money, either cost must be lowered or the perceived quality (being either tangible or intangible quality) must be raised. One method of raising intangible quality that is becoming more prevalent within building services is the use of brands.

Rethinking Construction, picked up on this idea and stated that:
"*...too many clients are discriminating and still equate price with cost, selecting designers and constructors almost exclusively on the basis of tendered price*". And that "*the industry needs to educate and help its clients to differentiate between best value and lowest price.*".

Clearly branded services and systems would even allow occasional clients to distinguish between features other than price. Currently occasional buyers of building services are unfamiliar with the technical quality aspects, thereby relying on such issues as price to decide value. This price-only approach reduces the construction industry to the role of commodity provider. There is little profit or customer delight in a commodity purchase.

It is important to clearly understand what is a brand. It is more than just a logo, but clear method in which a company carries out its business. One definition that could be used is:
"*enables its consumers to distinguish its products (or services) from those of its competitors...so the name comes to symbolise more and more about the sorts of products they can expect, above the level and nature of the service they will receive and even then above the working practices underlying these outcomes.*".

Through the use of brands, customers begin to expect a certain level of quality, resulting in a pseudo-standardisation.

Few people, however, especially one-off procurers, would recognise overall building brands. Likewise if the analogy of computers is used, few people are used to recognising brands within computers. However, Intel with their trademark of "Intel Inside" has shown how one can brand a hidden product. It comforts the consumer by ensuring that quality will be delivered, because of Intel's brand quality. Similarly, occasional construction procurers might be comforted by the brand of "It's got Company X air conditioning inside".

Brands can also create customer loyalty to both companies and products. Rethinking Construction states:
"*with few exceptions, investors cannot identify brands among companies to which they can attach future value... discussions with City analysts suggest that effective barriers to entry in the construction industry, together with structural changes that differentiated brands and improved companies' 'quality of earnings' (i.e. stability and predictability of margins), could result in higher share prices and more strategic shareholders.*".

The greatest potential for brands lies within the supply chain approach. Standardised systems that are normally delivered through a joint supply approach, such as air-conditioning systems, steel frames, custom metalwork or electrical installations. Clients have difficulty in procuring these types of systems as they are unable to distinguish the differing types of supply or organisations providing them. Arranging project teams around the product would enable brands to be developed in market sectors, finally enabling a building to be constructed in a modular approach by branded elements.

All of this results in the development of virtual companies, where an integrated supply chain forms around one perceived brand. Replacing the need to organise specific teams for each project with consistent teams based around products and providing a clear brand identity would have the double advantage of reducing problems and increasing value.

The development of brands would also affect procurement. Companies that possess strong brand identity normally control the manner in which they are procured. Customers would be freed from the traditionally complex arrangements and would be free to procure complete installations direct from the manufacturer or from a virtual company offering a multi-manufacturer solution.

Rethinking Construction also states:
"The repeated selection of new teams in our view inhibits learning innovation and development of skilled and experienced teams. Critically, it has prevented the industry from developing products and an identity - or brand - that can be understood by its clients."

Rethinking Construction rightly points out, over 80% of inputs into buildings are repeated. If a brand strategy could offered quality assurance would become immediately inherent with over 80% of the product.

Standardisation through brand strategies would provide the following benefits:

- reduced fragmentation, through a supply chain strategy of standardised product attributes, requiring joint working and innovations in product development;
- a focus on the product through integrated processes;
- focusing on the customers value hierarchy;
- identifiable quality benchmarks for occasional and inexperienced clients;
- the transfer of competition from lowest cost to value-based attributes.

This issue is particularly important to building services because of the current reliance on bespoke designs for service delivery. By forming a product around a brand, the client has a simplified procurement strategy available. Rather than face the complexities of assembling a team and executing work, the client is able to assess the competence of one solution compared to another. The result should be enhanced value and simplified procurement.

5.3 COST STRATEGY

Although progress has been made in recent years to refocus the decisions mechanisms within procurement from merely time, cost and quality to more dynamic criteria, whole life cost remains largely unconsidered. Building services are dynamic entities that consume money, even when laying idle. To maintain their value and usability, routine maintenance must be undertaken. Furthermore, the capital installation cost is only a small portion of the entire cost. Energy consumption, routine maintenance, operation strategies and climatic conditions all influence the total cost of the services, some of which are beyond the control of any party.

The costs associated with building services can be divided into two distinct categories:

- **capital costs:** all costs incurred with the initial design, procurement, installation and commissioning. These can range from 15% of the total project cost for a simple industrial unit, to 80% for complex medical and computer suite installations.
- **life cycle costs:** all costs involved with operation, maintenance, interim replacement and upgrades, and decommissioning. Energy consumption will be the largest element of this cost. Although subjective, it is normally considered that up to 90% of the running cost of the building is attributable to the operation of the building services.

It is within the briefing stages of procurement that clear policies and budgets must be determined from the client as to the balance needed between the costs. Differing client motives will alter the balance.

Developer clients building for resale only, will only consider capital costs, as the purchaser will incur the running costs. To maximise profit the developer must build for the lowest capital cost, although this trend is gradually being challenged. Greater awareness by purchasers and consultation with their facilities management staff has resulted in a more critical approach when reviewing the purchase of a building.

For most other types of clients the overall cost of operating the system and financing the initial capital installation are the most important of concerns.

Value is a relative concept. It should be judged on the basis of factors that are critical to the success of the project, through the eyes of the client. These are called critical success factors. The critical success factors of a project differ from one client to another. For some clients low capital costs are more important than low life cycle costs. Some may require a building designed and constructed in a short space of time in spite of the high capital costs it may cause. In each case the client may achieve their value criteria with different construction solutions and varying costs.

This requirement for control over cost can often cause problems within the development of a procurement strategy. As most main project strategies are based upon initial capital cost, the client is often left with a system that is disproportionately expensive to operate and maintain. It is for this single reason that the procurement of building services must be carefully considered prior to the main project strategy.

5.3.1 Capital Cost

The success with which capital costs are controlled during any building services installation will be subject to the coherence of the building services team. This team is an amalgam of designers, engineers, component suppliers, specialist contractors and the client's own representation. Each member has their own objectives for participating within the project, thereby affecting costs, together with the overall project environment within which the team must operate.

As the building envelope poses an effectual relationship over the building services, the procurement strategy for the project must be co-ordinated with that of the building services. All decisions made for the building envelope and structure will affect the design of the services, and therefore their cost. This relationship does cause problems between the project team and the building services team. To overcome these, cost expertise must be available within the building services team through either a dedicated resource, such as a cost engineer or quantity surveyor, or through the estimating expertise held by the services contractors.

The main cost drivers behind the project must be identified and incorporated within both the procurement strategy and the design brief. Generally for the building services element of most projects, cost drivers can be identified as coming from one of three sources:

Project drivers include:

- client needs - technical, financial and operational
- time available for construction (quick construction incurs cost penalties)
- the nature of the work
- maintenance strategy
- site conditions
- climatic conditions.

Procurement drivers include:

- the supply chain strategy and depth to which it is engaged
- the allocation of design responsibilities
- the price/quality mechanism to be used
- contractual arrangements
- market conditions.

Technology drivers include:

- construction quality
- innovation in design and installation
- construction efficiency and productivity
- the nature of materials and components
- the regulatory framework.

Within the context of procurement the greatest factor that will affect the capital cost is the extent to which the supply chain is engaged with both the design development and component supply. Although this subject is discussed in depth later in the text, at this point it must be understood that the more direct the relationship between the building services team and the component supplier, the lower the cost of supply. Each layer of subcontractor or subsupplier adds a percentage for overheads, profits and handling. The more efficient the chain, the lower the cost of supply.

Beyond its strategic nature, a number of operational factors will affect the cost of material and plant supply. In addition to the natural laws of supply and demand that affect the market place, most of the others are based on the end design which determines the overall size of the project. Such factors include:

- the size of the order placed with an individual company
- the discounting norms of the industry or company
- the nature of the component
- quality level
- after-sales support
- product brand.

The range of markets now available to supply components means that in fact most components suppliers now operate on a global basis; even the simplest of components can be selected from a bewildering range. Even when faced with identical quality standards, a wide range of supply costs are possible.

These variations are normally due to the purchasing nature of the contractor, coupled with the supply methods employed by the manufacturer. In addition to these, other factors include product features namely, for example:

- design duties
- functions
- maintenance requirements
- adaptability

and market features:

- the country of origin
- delivery logistics
- EU import restrictions
- product testing and certification
- the purchasing strategy adopted
- the marketing or supply strategies adopted by the manufacturer.

It is improper to suggest that only capital costs should be considered at tender stage. Capital construction costs are a small but significant part in a range of complex financing calculations required to bring any project to fruition. What should be aimed at in the early stages of the project is highlighting the interaction between capital and whole life cost. During the development of the procurement strategy a clear statement of objectives for both capital and whole life cost should be obtained from the end client.

5.3.2 Whole Life Cost

The simplest of building services installations will consume up to 10 times their installation cost in operational and energy costs, therefore whole life cost provides a more apt method of understanding the financial implications of a given design.

The whole life cost of an installation is the total sum of the initial capital cost, finance cost, operational cost, maintenance cost and eventual disposal cost, reduced to a present level amount.

Whole life cost is a misnomer, as it is a collection of several management techniques that use cost as a common denominator for ready comparisons. It considers several factors, including maintenance, functionality, risk and cost over the lifetime of a product, facility or system.

The basic definition can be used to further elaborate, depending upon the objectives of the party concerned.

- for designers it is:
  "an economic assessment of competing design alternatives, considering all significant costs of ownership over the economic life of each alternative, expressed in equivalent monetary value"
- for clients it is:
  " an economic assessment of an item, area, system or facility, considering all the significant cost of ownership over its economic life".

The idea that whole life costing extends beyond purely economic considerations was developed in the 1960s with the establishment of terotechnology, which is defined as "a combination of management, financial, engineering and other practices applied to physical assets in pursuit of economic life cycle costs".

The current move towards outsourcing and non-capital commercial financing has created a demand for purchasing decisions being made over a lifetime by both supplier and procurer. It is also being driven by the exorbitant cost of a building project, which given the shortening life cycles of products/services and companies, means that all buildings severally affect the business case, and therefore must be economically viable themselves as a business function.

Basic Principles

Whole life cycle costing is a technique currently used for the comparison of different investment options on the basis of the assessment of their capital and future costs / benefits over a period of time. It is commonly used for:

- the design process for the economic appraisal of different plant and installation options
- assessing the operational life to arrive at repair or replace decisions.

To adequately undertake this analysis requires the establishment of three key factors: time, defined as the service life; cost, discounted to the present; and the underlying economic principles.

Present value is used as the basic monetary principle because of the erosive effect of time on the value of money. It is a well understood principle that the value of money erodes over time. For example, consider three possible alternatives: is it preferable to be offered £100 now, £100 in a year's time or £10 per month for ten months? Which is the most desirable alternative? If a three per cent inflation rate is assumed, then present value would discount the amount to £100, £97 and £98.80 respectively. Not only must time be considered, but opportunity costs and the risk of collecting the amount over extended time periods.

Cost is merely the common denominator of a series of decisions that must be made. Risk management is used within the procedure to allow improved decision making using either a sensitivity or probability technique.

Using whole life costing within the design stage allows the proper identification and evaluation of the cost of the installation. Any design decision should be made with regard to entire cost, as low capital cost, for example, may cause high running costs. For any decision made at the concept design stage will affect running costs in the future and the overall economic life of the building. However, whole life cost does justify good engineering. Improved design may cost more initially, but justification can be made for it by reduced operating or maintenance costs.

Whole life cost management uses the costs identified in life cycle cost analysis to identify and minimise cost expenditure in the operation of a particular item of plant. In addition to design, it can be used as a decision tool in four ways:

1. As a total cost commitment to an item rather than initial capital cost;
2. By allowing a decision to be made between two or more options, using alternative methods to achieve a stated objective;
3. As a management tool that details operating cost;
4. By identifying areas where operating costs may be reduced.

Net Present Value

The purpose of the net present value model is to optimise the return from an investment and is suitable when the plant life and the running costs/savings are fairly certain over a period of time. The model works on the principle that it is desirable to recover the cost of the investment at the earliest possible time. This is particularly appropriate when there is the risk of short term obsolescence or if there is a risk of significant variations in costs/savings in the long term.

The basic model is expressed in the formula:

$NPV = X/(1+r)^t$ where:

NPV = net present value
X = initial capital sum
r = rate of interest
t = period of time

This formula can be extended and adapted to provide for multiple discount factors. A full explanation of the formula is given in Chapter 8.

Where the data and assumptions used have a high degree of uncertainty it is prudent to carry out a sensitivity analysis to establish whether the outcome from a costing exercise significantly changes under different scenarios of costs and assumptions that may be possible.

Whole life costs can be limited in scope if it does not take account of the effect the installation has on the client's business operations. Therefore, whole life cost comparisons should also take account of possible business costs due to reasons such as plant malfunction and service obsolescence which could be equally, if not more, vital to the client operations.

Other costs that must be considered over the lifetime of a system or building include:

- **Financing costs:** to finance the debt caused by the construction, often very complex calculations that take the form of a separate business case.
- **Operation costs:** the costs associated with the running of the system.
- **Maintenance costs:** for regular custodial care and repair, annual maintenance contracts, and costs associated with the facilities management function. They normally consider replacement costs of a product on less than 5-year cycles, or of a relatively small monetary value, i.e. £5k (larger items are considered as single replacement cost items).
- **Alteration/replacement costs:** usually involve large scale replacement or change of function for a space. Replacement costs are items that must be replaced during the design timeline due to their inability to survive for the total time period. For example, the best cooling towers available will only have a useful life of 20 years, therefore to suit a thirty year timeline, will require two units or a comprehensive maintenance programme (together with a careful initial design) to maximise the life.
- **Tax elements:** consideration of all the various taxes and credits may affect the initial decisions over both cost and systems design. These may include carbon tax, capital gains tax, tax relief and capital allowances.
- **Associated costs:** a mixed bag of "catch all" costs associated with functional use, denial of use (lane/road/track rentals?), security and insurance. Denial costs could be extreme. A recent example is that of a major London bank headquarters (incorporating a trading floor) that must be closed yearly to service the cooling towers. A two-day shut down causes a loss of business valued at £500k. Careful consideration at the design stage should have prevented such a cost, through the selection of an alternative system, or duplication of towers.
- **Salvage value:** may be a cost or value depending on the final economic worth of the system. Normally buildings would have a positive value, but few examples of services can be given, except for major plant items still capable of economic functioning, e.g. boilers, large air-handling units or containerised plant rooms.

Determining Service Life

Service life planning is a design process which seeks to ensure, as far as possible, that the service life of a building will equal or exceed its design life, while taking into account (and preferably optimising) the life cycle costs of the building.

Although a range of methods can be used, ISO 15686 provides an internationally accepted methodology for forecasting service life and estimating the timing of necessary maintenance and replacement of components.

The standard utilises a prediction method for long-term performance based on exposure and performance evaluation. A reference service life is used, then adjusted using various adjustment factors:

- factor A: quality of components
- factor B: design level
- factor C: work execution level
- factor D: indoor environment
- factor E: outdoor environment
- factor F: in-use conditions
- factor G: maintenance level.

Determining the service life of an installation must be carried out on two differing levels. Firstly, the overall life cycle of the installation must be determined. This is normally assumed to be 25 years, but will be dictated by the business objectives of the client. The second calculation is the more complex part. Each component must be assessed as to its probable or desirable service life. For example:

- **source units:** most modern sources units, e.g. boilers, are capable of operating for 25 years and are normally too large and complex for an installation to consider in short time periods.
- **distribution:** pipework and ductwork would normally be divided into primary and secondary installations. Primary pipework is often buried within the building and is normally designed for the 25 year maximum life. Secondary pipework is normally subject to the alterations of interior fit-outs and interim plant replacement. Therefore its design life should be equal to or greater than the expected interior life.
- **outlets:** items such as radiators are often replaced with interior alterations. Careful assessment of the client's facilities management strategy and approach to interiors should be made.

Service life planning aims to reduce the costs of building ownership. Assessment of how long each part of the building will last helps to decide the appropriate specification and detailing. When component and system lives have been estimated, appropriate maintenance planning and value engineering techniques can be applied.

The objective of service life planning is to ensure that the estimated service life of the building or component will be at least as long as its design life. The purpose for most projects will be to ensure that the most advantageous combination of capital, maintenance and operational costs is achieved over the life of the building. The output of service life planning will be a series of predicted service lives of components, and a projection of maintenance and replacement needs and timings. Achieving this will, of course, require maintenance during the service life of the system and its components.

Thus estimation of service life at the design stage assists the planning of future maintenance operations, selection of the optimum specification and design, and the avoidance of waste.

....

Case Study

MOD Prime Contracting

The MOD Prime Contracting procurement initiative is being driven by the Building Down Barriers initiative. It was set up by the Ministry of Defence, to deliver three overall objectives:

- to develop a new approach to construction procurement based on supply chain management,
- to demonstrate the benefits of the new approach, in terms of improved value for the client and profitability for the supply chain,
- to assess the relevance of the new approach to the wider UK construction industry.

Whole life costing plays a major part in this strategy. It is a requirement that all designs be based on the costing, including relevant use of risk and value management techniques within the design phase. In the words of the MOD "*Make value explicit: design to meet a functional requirement for a through life cost.*".

Completed projects are assessed for the first two years of their life as to the accuracy of the developed costs. Given that this is a single point responsibility procurement arrangement, consultants and contractors are collectively held responsible for this accuracy.

It is interesting to note that studies undertaken by the Tavistock Institute during the two pilot projects showed no decrease in capital costs but significant savings in whole life costs. These were achieved with the introduction of key suppliers and clients within the detailed design stage.

From Tavistock's report, seven key principles have been determined that are critical to the success of this procurement arrangement. Two of these explicitly detail the requirements for whole life costing.

Principle 1: Compete through delivering superior underlying value rather than lower margins.

Principle 4: Make value explicit: design to meet a functional requirement for a through life cost.

Whilst these principles are easily understood, both have had trouble being established. The principle of using whole life costing is based upon the comparison of a cost model of the proposed design against an established Historic Reference Cost of previous MOD properties.

This process is supposed to lead to two major improvements within the construction process. The first is greater involvement of the supply chain to rid the construction process of inefficiencies within both the design and assembly operations. This should then allow savings in efficiency to be used for improvements in material specifications, process improvements, and buildability. The second improvement stems from the first in that the end project should have increased value and quality, leading to improved whole life costs.

The two pilot projects are on course to deliver cost savings of around 10% in both capital and whole life costs. Moreover, a greater level of satisfaction is enjoyed by the project stakeholders with the final design and its functionality.

....

5.3.3 Value Engineering

Value engineering is a systematic management tool used to ensure that money allocated to a project is spent on the client's actual needs, rather than on the perceived requirements of the design team. Uniquely within the US construction industry, it determines functional requirements rather than merely being used as a cost reduction aid.

Four definitions are required for the understanding of the basic principles:

- **value engineering:** a systematic approach to achieving the required project functions at least cost without detriment to quality, performance, and reliability.
- **value management:** a structured approach to defining what value means to a client in meeting a perceived need by defining and agreeing the project objectives and how they can be achieved.
- **unnecessary cost:** cost which provides neither use, life, quality, appearance, or customer features.
- **value analysis:** an organised approach to providing the necessary functions at the lowest cost, or, an organised approach to the identification and elimination of unnecessary cost.

The above definitions must be considered in the following context:

- use refers to the utility of the component
- the life of the component or material must be in balance with the life of the assembly into which it is incorporated
- quality is a subjective function
- product appearance is often one of the most important client features
- customer features are ones that sell.

Value Analysis

The key point to value engineering is ensuring that the client's needs and wants, together with their inherent value, become properly apportioned within the overall project scheme.

The client's value system can be used to audit the following:

- the use of the facility in relation to corporate strategy
- the project brief
- the emerging design
- the production method.

It must be noted that value engineering is not merely a cost cutting or account auditing system. It tries to match the client's functional requirements with cost expenditure to ensure that adequate funds, in the correct proportion, are being spent on the right items. Value engineering and account auditing differ by the following main principles. Value engineering is:

- positive and pro-active by the use of a team-orientated creative process to generate alternatives to a stated solution,
- a method that is systematically structured with an established relationship between function and value.

Functional Analysis

A functional analysis sets out the "how" and "why" requirements of an element or complete project in a linear diagram. The diagram is normally referred to as the functional analysis system technique (FAST). The main principle of this technique is to initiate a process to attain the most appropriate conception and development of a system to match the user needs in terms of quality/cost/time ratios.

FAST diagrams are written from left to right, moving from the main project objective to the lower level ones. Also, the primary functions or requirements are stated as the main project objectives, this then leads on to secondary ones. In order to construct a diagram several basic rules are followed, namely:

- verb/noun definitions: the function of an item is expressed with one word
- technical solution: these are represented by a component or element
- function definitions: these are defined after the technical solution is found
- primary solutions: are those without which the project would fail or the task could not be accomplished
- secondary functions: these are characteristic of the technical solution chosen for the primary function.

Analysis begins with asking the question: "what does this do?", which leads on to the value management question of: "how else can this be achieved?". The answers to these questions and details can be combined into a function specification. It is critical that the precise function of each system is determined. This will help eliminate any functional redundancy and ensure that only the client's requirements are fulfilled as effectively as possible.

As well as defining the project's true functional requirements, additional objectives of a detailed function specification include:

- to help the company set goals relative to the product
- to separate the responsibilities of user, supplier and designer
- to define functions both qualitatively and quantitatively

- to set up a grid for the analysis of solutions.

Cost Reduction

The most important aspect of value management is in determining functional analysis over and above cost reduction. Cost reduction is a secondary feature and the by-product of the functional analysis exercise.

The construction of a new building requires the outlay of vast capital and considerable resource commitment. The resulting buildings function will directly determine the success of the business and its implementation strategy for expansion. These same concerns can be expressed within the commitment to a manufacturing set-up. However, the possible success of the building cannot be easily tested or modelled. This is where value engineering can help, in the minimisation of possible negative impacts, by ensuring the utility of the building suits the user's requirements without undue excess.

Value engineering is more than a double check of what is being proposed by a design team. It allows a proactive review of the proposals and the checking of alignment with the company's own business strategies. It must be acknowledged that most design teams have vested interests in the design and construction of any building. An independent value expert ensures the client's needs are being met together with their maximum number of wants.

Value engineering offers a well developed management tool for the strategic reduction in project or element cost, without sacrificing the client's end desires. It particularly lends itself to building services because of its concentration on the element function being matched with a value hierarchy.

To become successful and widespread in the services industry value engineering must be adopted and implemented within the normal duties of a building services engineer. New skills are required including whole life costing, briefing, risk management and a more detailed understanding of value engineering than given here.

5.4 PRODUCT STRATEGY

5.4 .1 Standardisation

Standardisation of processes and the product delivers a number of benefits, from the overall efficiency of repeating operations to the effective management of spare parts.

Standardisation must be outlined within typical business terms. This must be done to meet the call for reductions in capital cost and improvements in project delivery - in terms of cost, time and quality. To do so requires the exploration of certain slightly abstract principles, such as its relationship with quality and product-based supply strategies.

The Role of Quality

The first question that must be asked is whether standardisation is aligned with quality. Quality is about meeting the customer's requirements. The acceptability of a product, and hence customer satisfaction, depends on its ability to function satisfactorily over a period of time. This could be defined as reliability. Reliability is the ability of the product to continue to meet the customer's needs. Standardisation provides the continuance of the service, as the product performance should remain the same every time. Therefore quality equals standardisation.

Quality of design is determined by the ability of the product to achieve the agreed requirements. But do all customers want the same thing? This gives the determinant of the level of standardisation, as to whether it should be the entire product or for components only. What is received by the customer should conform to their requirements and hence to the design. Operating costs are always linked to the level of conformance required.

Critical Issues

Standardisation has thus been outlined in general terms, including some of the conflicts of interest that arise. In principle, four key areas tackle the subject in more depth, although obviously not all are relevant to every situation.

Process

A process can be defined as the transformation of a set of inputs, including actions, methods and operations, into outputs (products or services) that meets the client's needs. For building services this could include the function of the end system. To produce an output that meets the client's needs, consistency in the inputs are required. This eases the level of control required to define, measure, monitor and manage the quality of inputs.

Quality control involves key processes being monitored to ensure successful completion. Standardisation of processes and procedures, however, allows the removal of quality control - it becomes inherent within the tasks. It also aids in the production process as the emphasis transfers from designed solutions to technical and process led ones. Reduced design variations allows a quicker and lower risk response in construction.

The number of interfaces are reduced as the technology emphasis is narrowed, so the results are fewer planning and control problems. Functional analysis can also be reduced during the design phase as predetermined solutions can be selected.

Standardisation of work processes - systems that specify how work should be done – allows the movement away from a heuristic craft function to an optimised task plan working. Standardisation of outputs, or service level agreements, is used to demonstrate that work outputs will meet a given standardised delivery. Mutual tasks can then be co-ordinated around a given set of output parameters. All of this can lead to the standardisation of skills, including the sharing of mutual knowledge. Multi-skilling can only be achieved through the sharing of core skills and knowledge.

Product

Standardisation minimises the number of interfaces across which communication is required. Construction and its products are divergent in nature. The overall number of components and arrangements are becoming wider in range as the industry ages. While divergence is attractive to engineers and architects, allowing them considerable freedom in the development of unique designs, it is detrimental to standardisation on a product basis.

Convergence, being the narrowing of design options through process optimisation, is needed to achieve standardisation and its inherent benefits. Rationalisation can be introduced to minimise the number of alternative products or processes utilised. Its use results in a change in the construction process due to a self-imposed limited choice as to the number of materials, components, sub-assemblies, skills and plant permissible.

Therefore rationalisation can be used to enforce standardisation. Resistance to standardisation extends from the designer's perception of the loss of creativity. Creativity is seen as the balance between the divergence and convergence of extremes. However, building services' primary requirement is to function. Few components act in an aesthetic manner. Therefore the benefits of standardisation should assist at least in meeting a client's product function requirements.

Innovation

Innovation entails the conversion of radically new ideas into products, and the continuous development of existing products to increase their performance and alignment with customer requirements. Innovation plays an important part in building services, allowing for a reduction of capital and life cycle costs.

Either the market or technology can drive innovation. Currently building services is driven by technology on the perception of client needs. Standardisation requires markets driven by an innovation culture when standardised components are assembled into products that meet the client's exact needs.

Currently the construction industry is able to respond to each client's demands in a particular fashion. Of all industries, construction provides the ultimate in flexible manufacturing. Standardisation, while allowing increased opportunities and linking with client's requirements over conformity in systems assemblies, will not allow the direct matching of requirements with products.

Specifications

Specifications can be used to deliver standardised products. A standardised product is something that can be replicated over and over again giving the exact same outputs. Specifications must therefore be written in a manner that allows this to happen.

Platforms can also be used, in that specifications are modular allowing certain standard elements to be developed. National or international standards can be implemented allowing criteria on performance, safety or technology to be set. Innovation and product differentiation would come from the unique assembly of modules.

Standardisation does not, however, allow the best design or specification to be utilised, merely the closest match. This is a breakaway from traditional engineering teaching where the optimum solution is sought. Furthermore, it may actually slow down the pace of innovation.

A balance must therefore be struck between innovation and standardisation. In the interest of performance and safety, service systems are normally based on past solutions, utilising past proven components. What is required is a balance between materials, products and processes, against a background of known reproducibility and reliability.

The Role in Procurement

Currently, building services procurement is a disjointed process between all parties within the supply chain. Clients desire to procure building services in a manner that matches their basic description of the functioning service, i.e. clients normally express their desires as the basic function of the product in terms of its basic outputs - heat, coolness, light etc. Their description is in line with that of a single product.

The aligning of building services to that of product strategies would facilitate the increased uptake of manufacturing strategies utilising standardisation, modularisation and preassembly. A greater capacity for bearing risk and increasing value would be a by-product of this type of strategy. Individual companies could come together and form themselves into virtual networks.

The disjointed working of building services causes the client to become disgruntled as a circle of blame exists between all project participants. With the use of product-based strategies clients would have simplified lines of responsibilities, together with integrated warranties and product guarantees.

Ideally, a system should exist that lays out the project requirement on a function tree. Each building element could then be assessed against a standardised element for compatibility. Partial solutions or modifications to standardisation processes may also be considered. This should also readily lend any project to a supply chain strategy as the product function can easily be seen on a multi-layered basis, i.e. both a large module and component strategies can be developed.

This should also allow the identification of the necessary attributes and functions of a building service's installation, on an elemental basis. By identifying the attributes and functions a strategy can be developed by industry to effect the possibility or optimisation of the supply of functions. Simplistic buildings may lend themselves to a single product strategy, whereas complex projects, such a hospitals, may more readily lend themselves to a number of product lines. This would assist industry in developing appropriate supply strategies of working together with other suppliers to deliver a product-based service.

The use of any strategy should concentrate on the business case as much as that of technical considerations. Current research is concentrating on the technical while ignoring the business case. Work by BSRIA shows that 44 out of 100 criteria for the success of manufacturing-based strategies are founded upon management decisions alone. Often the business case will immediately cancel out the practicality of standard solutions. The main issues surrounding the business case for standardisation may include:

- a clear understanding of needs
- aligning processes to manufacturing
- optimisation rather than maximisation
- the early involvement of suppliers (for knowledge input)
- interpretation of design and manufacture
- a suitable procurement arrangement.

The end result should be the development of buildings that are sustainable through component reusage and replacement, minimise resources through the elimination of functional redundancy, and develop a procurement methodology which is based upon function rather than exaggerated ideals. The end result will be standardised whole life costs, with the introduction of component reusage, thus extending the functional ability of any system.

5.4.2 Off-site Fabrication

Prefabrication has fallen in and out of favour several times since the Second World War. With it, so has the specialisation of techniques by contractors. Pre-fabrication hit its heyday in the 1960s, with the concentration of the entire industry on industrial construction techniques. Most contractors had specialist teams that performed prefabrication techniques on site, prior to them being craned into the main building construction. With the demise of large social housing projects and the move to specialised subcontract labour, prefabrication skills and techniques fell out of favour.

Major clients with repetitive build programmes, such as supermarkets, who were seeking improvements in construction times, revived these skills in the 1990s. With a growing number of major services contractors offering it as a service, together with the development of specialist prefabrication contractors, its use is once again gaining popularity.

Prefabrication and preassembly of building services offers the client and the building services team a number of advantages. These can include the three cornerstones of procurement:

- cost savings
- improved quality
- reduced on-site programme times.

These savings are attributable to improvements in:

- faster return on investment
- improved and simplified site management
- less site labour and movement of materials
- space savings of installed services.

Although prefabrication can be used with any procurement route, three routes in particular characterise the main factors that are needed to encourage its use. These are:

- framework agreements that allow a continuous and timely involvement of the entire supply chain. Continuous involvement allows the clients needs to be assessed, then the development of a strategy to meet these needs over several projects,
- partnering agreements which allow a teamwork philosophy to develop where co-operation and focus on the end objective should encourage the necessary environment to allow the use of prefabrication,
- design and build which encourages the integration of design with fabrication, allowing a composite solution.

Although prefabrication is a technical subject, the procurement arrangement used for the service's installation will have a considerable impact on the ability to successfully implement a strategy. Of the barriers that are perceived to exist, those specific to procurement can be grouped into three main issues: the supply base and its necessary skill level, the design of the project, and the inter-relationships of the services team.

Research by BSRIA has shown that there are six major building blocks (Figure 5.2) for successful prefabrication.

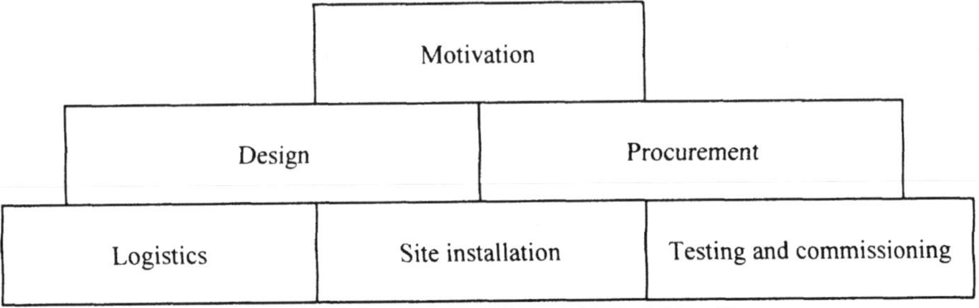

Figure 5.2 Building Blocks of Successful Prefabrication

During the developing of a procurement strategy, the decision on prefabrication will have to be made in order to secure a contractor and designer with the necessary expertise. A balance will have to be made between the services team's project expertise and their knowledge of prefabrication. The important consideration is to have a fully operational system, and therefore project expertise must dominate the decision-making process.

With this they must have sufficient capacity to carry out the work. Capacity can be augmented with a supply chain strategy. Specialist prefabrication works can be used to supplement the overall capacity of the lead contractor. Likewise, design and build product solutions, like complete boiler houses, can augment the design capacity. Overall the most important deciding factors will be the project mobilisation and delivery time, versus the available industry capacity.

The prefabrication contractor should not be selected on price alone, but on a range of factors including:

- in-house skills
- production capacity
- fabrication expertise – both managerially and of facilities
- a proven track record
- an appropriate approach to construction
- quality control systems
- design capacity
- delivery and scheduling logistics ability.

Project design will need to consider prefabrication at the earliest time, and incorporate its requirements within the overall project scheme. Ideally favouritism should be given to a procurement arrangement that combines design and construction. This enables the specification and detailed design to be developed with sufficient flexibility to enable prefabrication. To be successful the designer and contractor need to work co-operatively to achieve a fully co-ordinated detailed design.

Successful prefabrication, both technically and economically, relies on the repetition of elements. Therefore it is particularly ideal for projects such as hospitals, hotels and office buildings having large areas of common design. This limits the number of specific assemblies and aids in both tolerance and standardisation of components. Consideration also needs to be given to the testing and commissioning regime to be used.

Although prefabricated services are initially more expensive, the reductions in project time should generate sufficient costs savings to offset the increase. On a similar basis, the ability to fully co-ordinate and install a large item of prefabrication will involve increased risk to the services team. However, these are offset by the reduction in the number of site operatives required. To benefit, the services team needs to adopt a balanced approach to both the problems and benefits.

5.4.3 Service-based Agreements

Ever more pressure is being placed on commercial clients to reduce their capital costs. At the same time, suppliers of large capital items are seeking ways of generating a higher level of profit, without increases in the base cost. Most of these companies have known for some time that the real profit of an item rests with the continuous supply of spare parts and servicing of the item. By providing a good quality after sales service the prospect of the client returning for additional or replacement equipment is also raised.

This simple good business practice has been combined with sophisticated methods of financing and insuring businesses to create the service level agreement. First developed in the automotive sector under car leasing agreements, the general principles are now being applied to most capital equipment, including building services.

One of the big drivers behind service agreements is the power of some companies' brands. As described earlier in this chapter, the strength of certain brands provides immediate assurance of quality and service.

Taking the idea of design and build one stage further, the concept of a service-based agreement is born. Most companies are seeking three immediate deliverables from a service strategy – minimal capital cost, guaranteed running cost and minimal risk of failure. A service-based agreement provides part of a building services system of pay-per-use or on the basis of a set monthly charge basis, thereby minimising capital cost and providing an assured running cost. Risk is minimised as maintenance and periodic replacement of the system is the risk of the supplying company.

Although the allure of reduced capital costs may appear attractive to most clients there are several drawbacks with the idea. As the ownership of the equipment and system remains with the supplying company, the client cannot maintain, repair or modify the equipment. This can only be done by the supplying company or under their agreement. The client may therefore be forced into paying inflated costs for any modifications. In the same manner, the client is also reliant on the supplier to execute a proper maintenance strategy and remain in business throughout the life of the agreement. Any failure in business of either party would have serious repercussions.

The introduction of service agreements could occur with most forms of procurement, given sufficient time to write the document and sort out the contractual implications. However, they are best used where the client is both experienced and able to directly influence the supply chain. The contractual relationship will always be between the client and the supplying company, or possibly the main installation contractor. Therefore this relationship should be established in the initial stages of the project.

....

Case Study

Plant Energy Systems are prefabricators of boilerhouses and plantrooms. They provide purpose built rooms with a solid steel frame and clad with colour coated architectural composite panels.

Physical sizes vary from small 3 x 2 metre boiler rooms to vast energy centres built up in 18 x 4 metre modules. Duties range from 100 kW to 40 MW and above. The packaged plantroom contains all of the items normally found in any modern plantroom. The big advantage is that these can be assembled away from site, divorced from the problems associated with a busy construction site.

The advantages provided to a client are many, as they are designed as permanent solutions with life expectancies in excess of 25 years. However, they do have the flexibility that they can be moved at any time, should the needs of the building they serve ever change.

This concept combines several ideas: prefabrication, brand strategies and service-based agreement. Since their original inception as merely prefabricated plant rooms, Plant Energy Systems have reinforced the perceived product quality by openly marketing into products using their component supplier's brand names. Furthermore, they now offer the complete plant room on a service basis. This has been facilitated through a partnership approach with a major utility financing company that specialises in structuring service finance.

....

REFERENCES

1. Flanagan, R. and Norman, G. (1983) *Life Cycle Costing for Construction*, London, RICS Publications.
2. Hayden, G.W. and Parsloe, C. (1996) *Value Engineering in Building Services* 15/96, Bracknell, BSRIA Publications.
3. HMSO (1998) *Rethinking Construction,* London, DETR.
4. Johnson, G. and Scholes, K. (1993) *Exploring Corporate Strategy*, Hemel Hempstead, Prentice Hall International (UK) Ltd.
5. Kelly, J. and Male, S. (1993) *Value Management in Design and Construction*, London, E&FN Spon.
6. Oakland, J. (1995) *Total Quality Management*, Oxford, Butterworth-Heinemann Limited.
7. Wilson, D.G., Smith, M.H. and Deal, J. (1999) *Prefabrication and Preassembly, ACT 1/99*, Bracknell, BSRIA Publications.

Chapter Six

Quality

6.1 PROCESS QUALITY

The word "quality" has a variety of meanings, especially within the context of construction where it can be defined from an aesthetic, functional or legal point of view. However, synonymous with any of these functions is reliability and excellence. As far as building services are concerned, quality is the sum of faultless building services, maintainable services, punctual delivery, value for money, ease of maintenance, reliable systems and being fit for purpose.

Within the procurement function two aspects of quality exist.

1. **Quality Assurance:** a rigid set of defined procedures that aim to insure conformance to a set of requirements. These are normally described by minimum standards within the project specification and contract. Often they are set as overall performance standards in which the building services systems must achieve.
2. **Total Quality Management:** a management philosophy which aims to produce a better performance from a whole project team and to result in better quality products and services, delivery and administration, which ultimately satisfy the clients functional and aesthetic requirements to a defined cost and completion time. For this to work the client has to accept the responsibility associated with being part of the project team.

The role of procurement is to form a collective body acting as a single organisation to execute a project. Whilst quality standards can be defined through the contract documentation and set standards, ensuring the management structure for a project is committed to the same principles is inherently more difficult.

6.1.1 The Need for Quality

Successful building services require a myriad of consultants, engineers, contractors and specialist suppliers to work as a single entity. Only when the team is fully integrated will the necessary technical information be successfully communicated. The management structure must reflect that of the services installation – a seamless system functioning automatically.

Achieving quality in construction has been dogged by the myopic view that individual participants were responsible for their own quality, rather than viewing quality as a holistic concept. For efficient building construction, an overall quality system must be implemented and appropriately interfaced between all parties involved. The system must consider the social, technical and political environments that surround any project. By doing so the result will be more effective planning, better design, smoother co-ordination of services, improved building process and quality, and increased cost effectiveness - all carried out under a management regime that seeks fewer delays and disruptions.

Each procurement method available for a project will have different implications for and impacts on, the relationships between the client, designers and contractors and on the performance of the project. Hence, the choice of an appropriate procurement path as an integrated part of quality attainment should also be carefully considered.

Total Quality Management can be used as a framework to ensure the objectives set by the client are incorporated within the management regime. To satisfy a client's needs, the traditional difficulties must be alleviated through greater care and attention to quality and management systems within the briefing, design and construction stages of the procurement of a services installation. Therefore modern management tools must be used to reform the management structure and attitude of the traditional approaches to services procurement.

For the better management of quality in building services, aspects of Total Quality Management should be incorporated within the procurement process. Its successful implementation can achieve better integration of the project participants, allowing for improved co-ordination of services that will deliver a project quicker, cheaper and with improved technical quality.

To fully satisfy the client's requirements, the procurement strategy must allow the execution of the design and construction processes, enabling the following:

- integration with other elements to form a singular system, that is properly planned and co-ordinated for a smooth installation;
- satisfactory performance in terms of environmental comfort and provision of optimal convenience to users;
- effective operation of efficiency in both maintenance and energy consumption;
- cost effectiveness in terms of life cycle costs; and
- a design with the necessary levels of flexibility, adaptability, workability, reliability, manageability and safety in mind.

Regardless of the stage in the process, everyone should seek to identity what their customer requires. This should also be coupled with the idea that everyone has a customer both within and outside an organisation. Paradoxically, it must never be forgotten that a construction project is about delivering a fully operational structure to a single end client. The focus of the project must always be on this end goal.

Given the inherent complexity of building services design, together with the relationship between the supplying parties, it is quite possible that things can go wrong if there is a lack of structured quality management for design and construction. Each action undertaken by a party has a knock-on effect on subsequent activities or events, e.g. inadequate design produces unsatisfactory co-ordination, installation and maintenance of services, thereby affecting the quality of the end building.

In the context of building services, the problems can be broadly divided into three categories, namely:

- **Technical problems:** these include inadequate briefs, inaccurate or inadequate details of design concepts, lack of integrated and co-ordinated designs, incomplete design information, mistakes or discrepancies, impractical or complicated designs, unclear responsibilities, poor installation and difficult maintenance.
- **Management problems:** including inadequate management of the team members, inadequate management of the design, inadequate management of the services installation and commissioning, inadequate management of the multi-organisation, poor communication and site supervision, and the lack of acceptance of professional/contractual responsibilities.
- **Human relationships:** these relate to the effectiveness of team working and co-operation. For example, entrenched adversarial relationships caused by the traditional contractual and management styles of the industry may exist with certain team members.

The procurement arrangement selected for a project will address the first two problem areas. Regarding the third, although during the initial selection stage principal participants can be reviewed for compatibility, it is a false ideal that totally compatible parties are to be found. In order to meet the other necessary technical, financial and managerial aims some compromise will always be evident in team member selection.

6.1.2 The Concept of Total Quality Management

Human issues are difficult to overcome as the state of legal independence between participants and their traditional methods of working together are often an obstacle to adopting a team environment. To improve working relationships, a cultural change away from the adversarial attitudes that participants have traditionally adopted is required. To achieve improvements in the management process requires the adoption of a specific management strategy. Total Quality Management (TQM) is one such framework that embodies the attributes required.

These attributes must reflect what would be required to produce the ideal services installation:

- **For design-related problems:** comprehensive client's brief; adequate resources; good design management; fully integrated design; co-ordinated design and installation; buildability; complete design information; clear design and construction responsibilities; and continuous design review, checking, and feedback for improvement.
- **For co-ordination problems:** clear responsibilities, adequate time and resources; fully co-ordinated design; adequate on-site co-ordination management; combined services drawings; proper management and supervision of installation; adequate participation of design team.
- **For management problems:** effective project management; team working spirit; effective communication and interaction between client and services team; documented procedures.
- **For procurement problems:** comprehensive procurement strategy; fair and appropriate risk distribution; contract selection; compatibility of parties; previous experience; strategy for attainment of quality.

To achieve a quality building services installation, an overall framework across the total process, assuring quality, becomes inherent in all levels and stages. In this respect, Total Quality Management (TQM) provides a good environment for achieving this objective. TQM is different from quality assurance in that it is a corporate philosophy rather than a reliance on rigid procedures. The entire approach can be distilled down into ten key points for adoption:

1. A commitment to long term improvement
2. A "right first time" culture based on zero defects and positive action
3. Training of all people to adopt the customer-supplier relationship
4. Total cost and value must be considered, not just initial capital cost
5. Recognition that improvements must be managed
6. Adapting to a modern team-based culture of supervision and training
7. Managing the process to encourage communications and teamwork
8. Eliminating friction, arbitrary goals and any barriers to improvement
9. Constantly educating the team on improvements
10. Developing a systematic approach to quality management.

TQM is a cyclical process of executing an action then analysing the effectiveness of the process. A realistic plan, systematically prepared, indicates that the various design and construction processes and requirements have been fully reasoned. By doing so the likelihood of co-ordination problems or mistakes due to inadequate design and construction management are abated.

The goal of all this is to seek reductions in the cost of providing quality. Considerable costs are incurred on all projects in correcting deviations from the stated requirements in design, construction and occupation of the building. Studies have shown that the cost of correcting problems, on average, is in the region of 12-15% of the project cost, whereas the cost of providing TQM is between 1-5%.

6.1.3 Problem Areas in Implementing TQM

Providing any project-wide management strategy is always difficult when the participants are a diversity of people, professions and roles. This multiplicity of participants can cause an assortment of problems in the successful implementation of TQM.

These difficulties can include:

- complex layers of internal and external customers and suppliers involved in a single project;
- differences in the social backgrounds and managerial styles of differing parties;
- contractual interdependence or lack of relationship between parties;
- diversity of constraints set by codes, standards and laws;
- lack of commitment by management to change the culture and practices of the company;
- a procurement strategy, or its execution, that has not fully considered the required levels of compatibility or interdependence.

To implement TQM successfully, the building services team needs to concentrate on re-engineering the entire service process to emphasise teamwork and partnering between participants. However, to accomplish this a procurement strategy and arrangement must be used that allows the necessary flexibility and co-operation.

6.1.4 Quality and Procurement

The issue here is whether the quality of the delivered product is affected by the procurement arrangement. Two differences in opinion exist. One belief does not consider the impact of a contracting arrangement on quality to be critical, although in particular circumstances one arrangement may offer advantages over another. For example, highly serviced buildings, such as hospitals, may benefit from arrangements where the services contractor can have a direct relationship with the client.

The other camp views this issue differently as they consider procurement has a profound influence upon quality and therefore that the procurement system must be carefully chosen.

To achieve quality, a strategy must be devised that matches the procurement arrangement with an appropriate management style. Modern management belief is that teamwork should be treated as a first priority, thereby favouring non-traditional procurement methods. Caution should be exercised before automatically assuming that arrangements such as design and build or construction management may deliver superior levels of quality. Each arrangement delivers and controls quality in a difference manner. A detailed assessment needs to be made, rather than automatic conclusions.

What is important, is whether the selected arrangements and management approach allow for a good team-working spirit amongst the professional designers and contractors. Beyond the procurement arrangement itself, the manner in which the arrangement is executed will affect quality. The following issues commonly arise:

- The nomination process, when used in conjunction with traditional arrangements, can produce very good quality in building services. This is mainly due to the formation of a close relationship between designer and specialist contractor during the design stage.
- Regardless of the procurement arrangement, strong management and team-working are the most important contributory factors in achieving quality. This must be coupled to clear vision of what is to be achieved and how its quality will be assessed.
- An adversarial approach, which is often evident in traditional arrangements, will seriously compromise the quality of both the process and the product.
- Contractual arrangements have a significant effect on the working of the building team in terms of contractual relationship, working relationship, and the management of design and construction processes.
- The procurement arrangement and team members must be carefully selected to achieve the best quality attainable. Team members will come with predetermined attitudes, (they are only human), therefore effort should be made during the selection phase not to understand the company, but the specific individuals who will be undertaking the project.

The most plausible answer to the argument is that the various procurement methods available have different implications on quality, but that although the choice of contracting arrangements is an integrated part of quality attainment, it is not the determining factor. Quality is about attitude. Tools such as TQM provide a frame of mind that allows the necessary communication and commitment to achieve the desired level of service and quality.

With building services, quality is based upon a number of conflicting technical and subjective factors. A building services system may operate to technical perfection, but the occupants of the environment will judge its "quality". On a similar vein, the quality of a contractor or designer will be assessed against the subjective opinions contained within the project environment, and viewed by the client.

Effective management of design and construction will rely on good team-working which is an ingredient of TQM. Therefore the necessary level of quality should be inherent in the procurement strategy, since any one-procurement strategy would have a direct bearing on the performance of the client, designers and contractors. For it is the procurement arrangement that determines the particular project environment and characteristics.

A model can be developed that provides a better understanding as to how TQM and procurement arrangements can work together. The model encompasses functional requirements (i.e. sub-systems in the model) in terms of resources, then relates them to the attributes necessary to achieve quality.

These attributes are the essential ingredients of a procurement strategy and consist of time required for the management of design and construction processes, knowledge and competence of the design and construction teams, management issues, contractual matters and human factors. The resulting model is shown in Table 6.1.

Function	Quality-Procurement Attributes
Time Constraint	Sufficient time for the provision of a fully integrated and co-ordinated design, and complete design information Contractors' preparation of works, i.e. planning of all works, co-ordination of services and building construction Satisfactory execution of the whole project within a realistic contract period Good time management
Resources	Adequate funding for design and construction and realistic tender pricing Adequate resources for the preparation of the client's brief, co-ordination of building services and building construction by all parties involved Adequate management of resources
Competence and Professionalism	Competent design and construction participants of the right calibre and experience Quality of design and construction based on the client's need, with a TQM approach
Managerial Control	Effective design management of the integrated design team based on TQM Effective and flexible project management during the construction stage based on TQM Creation of team-working with good leadership and short - or long-term partnering relationship Effective and structured management of project information as covered in CIBSE's Quality Management System (AM9: 1993)
Contract Arrangement	Unambiguous contract, fair contract terms, clear responsibility, e.g., Engineering and Construction Contract Fair and clear risk allocation
Vision	Commitment to project success Trust and respect Team building for improving project performance

Source: Lam et al (1997) *Quality Building Services – Co-ordination from Brief to Occupation*, CIBSE Virtual Conference, London, UK

Table 6.1 Quality – Procurement Model

6.2 TECHNICAL QUALITY

6.2.1 Quality Assurance

An integral part of all modern management philosophies is that of continuous quality improvement. Where TQM concentrates on the soft issues of the relationship, quality assurance provides the detailed mechanisms and procedures to ensure compliance with the stated requirements. The two major elements of quality are quality control, which is the systematic assurance that the work produced complies with the set requirements, and quality assurance which provides the confidence to the receiver of the product or service that compliance has been achieved. The construction industry, in recent years, has begun to accept this notion and many have achieved quality assurance status. However, there is still no procurement methodology that tries to achieve this notion throughout the entire construction process.

Quality assurance must permeate the entire procurement process, from the initial client design brief, through detailed design to the entire manufacturing and installation process. Therefore quality assurance is a logical and progressive step, within the general management chain. Building services, given the requirements for logical and systematic approaches to their assembly, readily lend themselves to quality procedures.

Although building services contractors and designers have lead the field in formal recognition of their quality systems, there is a constant demand by clients to improve quality standards. In recent years most clients have demanded assurance to such standards as BS 5750 or ISO 9000. With the increasing emphasis on sustainability, most organisations seek assurance to ISO 14000 Environmental Systems.

Increasing, complexity in building services has only highlighted the need for greater assurance of quality, as greater complexity often leads to higher maintenance cost and incidences of failure. Therefore if quality and business continuity is paramount then the system should be designed as simply as possible. Complexity is often generated by sheer numbers and unworkable interrelationships between individual components. The manufacture of such components must be in line with the quality strategy, but few project quality strategies recognise this. Most fail on the following points.

- Little regard is given to protection of equipment during installation. Consideration also needs to be given to its accurate installation, protection during construction and exposure to external environments.
- The design engineer must precisely set performance characteristics. The manner in which manufacturers express performance varies from one to another, particularly if different test procedures have been used to produce the information.
- Initial installation is only a small element in the life cycle of a component. Assurance must be sought on operational, maintenance and fault detection factors.

- Quality can also be expressed in economic terms, especially in operating costs. Manufacturers must provide accurate information on operational costs and energy usage, allowing for the accurate establishment of whole life costs.
- Equipment life predictions are important in making cost in use calculations over extended timeframes. Assistance should be provided on calculations of useful or economic life, together with policies on spares and service availability.

....

Case Study

The Construction Quality Forum have published a guide to achieving quality in the procurement and construction of a building services installation. Entitled *Planning and Procuring a Building Services Installation*, the guide offers simple and concise watch points for each stage of a project.

During the design stage the report gives the following watch points:

- Ensure that persons having future involvement with the installation make an input to the design development process. These are to include operation, maintenance and safety staff
- The brief must clearly state all salient business needs to ensure they are considered within the design
- Critical questions should be made of the design including:
  - what level of failure (breakdown) can be tolerated
  - is standby plant required
  - can risk of failure be mitigated by hiring of plant or facilities in the event of breakdown
  - what level of risk in actual performance can be tolerated
- Make early decisions and monitor project progress closely
- Obtain buildability advice as early as possible during design.

....

6.2.2 Quality and Procurement

The benefit of using a standardised quality policy, such as ISO 9000, is that it provides a management framework for defining and meeting the client's requirements. Within manufacturing industries, where design and assembly are integrated, and procedures are rigidly set, the application of such a system is relatively straightforward. Referring such a rigid system to construction, however, proves difficult.

The manner in which construction is procured will affect quality. In keeping with the ideals of quality systems, the greater the concentration of activities, as in integrated design and construction, the simpler it should be to ensure quality. This holds true when the traditional procurement system is analysed. Separated responsibilities and individuals acting in independent roles, as in traditional procurement, provides for distinct problems in implementing a quality policy. The problems arise from the multiplicity of design and production responsibilities, reinforced by contractual arrangements that heighten the separation.

A coherent approach must therefore be developed in quality, but is often exacerbated by the conglomeration of performance, product and method specifications used within the procurement process. It is therefore important that a consistency in the specification requirements is evident within the contract documentation.

Beyond specifications, quality can only be achieved if all the diverse individuals within the building services team perceive the need for quality and demonstrate it through their actions. Following a disciplined approach, quality is achieved through the effective functioning and integration of clients, designers, suppliers and contractors. Each has a specific purpose within the team, which is determined by their function:

- **clients:** clear and precise definition of requirements
- **designers/engineers:** accurate translation of requirements into a functioning design and the specification of necessary components
- **suppliers:** delivery of properly functioning equipment
- **contractors:** physical construction and integration of all systems.

The key to achieving all of the above often lies in the clear definition of design responsibilities.

Roles and Responsibilities

In a simplistic vein, the process that generates a building services installation consists of two major elements, design and install.

The design element can be expanded to "design and specify", the objective being to define the responsibilities and procedures necessary to produce a design of adequate quality to suit the client's requirements, and to ensure this is clearly set down within the design drawings and specifications.

One of the major features of quality management is to ensure that equipment does not deteriorate during the installation process and it is essential that components be procured in a manner most likely to accommodate this. Logistical deliveries allowing for immediate installation are best. It is important then to identify clearly the procurement aspects associated with the installation process. Therefore installation should be interpreted as combining procurement and installation.

Commissioning, defined as setting to work and regulating, has traditionally been the responsibility of the services contractor, and therefore is normally considered as part of the installation process. Increasingly commissioning has become more formalised, with independent operators undertaking the role. There are good reasons for this development. Firstly, the commissioning process requires special skills and facilities, which many consultants or contractors could not afford to retain in-house. Secondly, the commissioning procedure validates both the design and installation aspects. If discrepancies become apparent, a neutral party, in the form of the commissioning engineer, is able to represent the client's best interests and work with all parties for a satisfactory conclusion.

Although not strictly part of the commissioning process as defined, this activity is often associated with testing to establish system performance, including the achievement of environmental conditions in the building. With the introduction of changes to Part L of the Building Regulations, this is now a set requirement. Quality of installation has now become a public issue with environmental performance being the centre stone for achieving it.

6.2.3 Client Role in Quality

Only the client can decide whether the level of quality delivered on a project is of an acceptable standard. And therefore the client should demonstrate their commitment to this level through the project structure, organisation and documentation. They can directly influence the project quality level by:

- appointing designers and contractors that have recent experience with the type of work, have a quality assurance scheme in place and have a managerial approach that embraces quality;
- insisting that components meet the respective quality approval systems, i.e. British Standards and/or Agréement Certificates, together with the more recent Energy Labelling Certificates;
- insisting that all major items of plant are independently tested prior to installation;
- ensuring that all sub-contractors and sub-consultants have equal quality standards as the principal contractor and designer.

On a more tactical level, clients can influence quality by addressing the key aspects of:

- **responsibility for quality:** who is to take the responsibility to ensure the level of quality required is actually delivered;
- **the quality system:** a formal system must be expressed in a documentation and fully co-ordinated to ensure consistent levels of quality;
- **planning:** there is a relationship between the time taken and the level of quality delivered. Project operations must be adequately timed and sequenced;
- **documentation:** the level of quality, together with performance and testing requirements, must be accurately stated in all documentation;
- **control:** consideration needs to be given as to how quality is to be inspected, measured, tested and controlled through the various parties;
- **commissioning / handover:** a formal procedure as to how the system is to be commissioned, including witnessing, and how the set quality standards are to be achieved, needs to be undertaken.

REFERENCES

1. Construction Quality Forum (1998) *Planning and Procuring a Building Services Installation*, London, CQF Special Report.
2. Lam K.C., Gibb A.G.F., and Sher, W.D. (1997) *Quality Building Services – Co-ordination from Brief to Occupation*, CIBSE Virtual Conference, London, UK.
3. Oakland, J.S. (1995) *Total Quality Management - the route to improving performance*, Oxford, Butterworth-Heinemann Ltd.
4. Watson, P. (2000) *Implementation of Total Quality Management in Construction*, Ascot, Chartered Institute of Building Construction Papers 110.

Chapter Seven

Retrofitting and Maintenance

7.1 WORK IN EXISTING BUILDINGS

It is estimated that over 85% of all contracts let involve the refurbishment of existing installations. Given the increasing pressures on building reuse, this figure is likely to increase.

Procuring building services containing refurbishment work involves the same general principles of strategy development, but two unique features will dominate the early stages of the project. Firstly, refurbishment by its own definition will involve the reuse of some existing plant and distribution systems. Therefore, the client's brief will normally be compromised by the attributes of the existing system. Secondly, before any judgements regarding design can be made, a survey of the existing installation must be made, together with decisions regarding the compromises willing to be made between the client's wishes and what the existing system can provide.

The above two features affect procurement. Procurement can only begin once a decisive strategy is developed for a project. The role of procurement is to take the strategy and break it down into a set of operational objectives. Refurbishment provides robust boundaries around most decisions: the strategy is partially dictated by the existing installation and structure. Therefore procurement must begin with an independent and objective assessment as to the available options.

7.1.1 Design Considerations

The extent to which the existing installation will be refurbished is largely determined by the change in use or updated requirements needed by the client. The business case for the project will set out the financial limitations of the extent to which things can be modified. However, further considerations must be made by the client with regard to future proofing the completed project from increases in technology, together with limiting the operating costs of the system.

Therefore, prior to undertaking a comprehensive survey of the installation, clients should consider whether to:

- take the opportunity to include energy-saving measures
- upgrade the technology
- allow for future capacity requirements
- replace plant with less noisy, more efficient equipment
- gain space through smaller or removal of redundant plant
- improve environmental characteristics
- improve control of the system
- reduce maintenance requirements.

Systems Survey and Evaluation

Within the context of procurement, the condition survey has three purposes, which are contingent upon the parties involved:

- **client:** To provide an objective evaluation of the existing installation through the establishment of a detailed asset list of existing equipment. This can then be used to provide details of existing plant performance and energy requirement which are useful for determining whole life costing calculations and capital budgets.
- **designer:** The schedules produced should form the basis of an initial options appraisal. Options can show what needs to be replaced or retained, thereby setting the general nature of the project. Combined with costing, full cost-benefit analysis can be undertaken. The resulting output must be a complete design or a schedule of works that can be used for a design and construct type of contract.
- **contractor:** The principal services contractor should receive the entire survey. This allows them to understand the nature of the project and the existing services, and to adequately provide for refurbishment and maintenance requirements. If the project is of a design and build nature, then the option assessments should also be provided, allowing the contractor to fully understand the decision process used.

7.2 MAINTENANCE

The single greatest issue that separates building services from all other elements of a construction project is that they are dynamic. While structural frames need little care or attention and interiors are generally short-term beautification, building services must operate in an efficient and effective manner. Not only must they deliver the necessary environmental conditions, but also to be considered successful, they must do so reliably and economically. To achieve this, and ensure the required health standards are maintained all systems require maintenance.

Therefore maintenance must be one cornerstone of the overall procurement strategy for a project.

In its most simplistic form three basic procurement options exist for maintenance: it is either ignored (which is outright irresponsible), becomes an extension of the installation contract, or is procured separately (often in conjunction with a facilities management contract).

Although a capital installation and maintenance contract can be combined, both are of a completely different nature, thereby dictating that either separate contracts or an overly complex one needs to be used. It is therefore advisable that a contract that combines capital and maintenance elements should ideally be a two-stage contract, whereby two separate contracts can be used, and providing a convenient break point for both parties should the first phase not prove favourable. Although a number of building services contractors offer both capital installations and maintenance, a two-stage methodology also allows the most appropriate contractor to be appointed for the work, as few contractors will posses equally favourable technical abilities and economical delivery for both types of work.

At handover building services systems are normally commissioned and expected to operate at peak performance. Without correct operation and maintenance the performance is likely to deteriorate. The installer's contract will normally include equipment warranty for a new building services installation but not maintenance. Inadequate maintenance may invalidate the warranty. Therefore, the issues must be considered prior to the appointment of the installing contractor.

The characteristics of maintenance works, together with the various technical considerations that must be made during the development of a maintenance programme are beyond the scope of this text. The following section concentrates on the general nature of maintenance contracts, thereby providing the reader with an insight into the considerations that should be undertaken when developing a procurement strategy. Detailed development of a maintenance strategy, although begun in the earliest stages of a project, is not fully considered until design development has begun and therefore after the procurement of the building services team. Therefore to fully develop a strategy requires an appreciation of the nature of maintenance contracts together with ensuring the buildings services team possess such capabilities as well.

7.2.1 The Marketplace

General construction is influenced by the traditional divisions of power and those held by the contractors and consultants. Maintenance is a modern industry that has not come under the influence of industry bodies or lobbying by biased organisations. Instead, the manner in which the supply chain is organised and service is provided are largely due to the size of market, influence of major clients and the characteristics of the supplying contractors. The maturing of facilities management into a profession has also had further influences.

The definition of procurement is extended somewhat when applied to buildings services maintenance. As the contract is normally directly between a client and contractor, there is deemed to be a certain level of knowledge and resources that are provided by the contractor to achieve a service level, rather than a defined output (like a building). Therefore for maintenance the preferred definition of procurement must be extended to incorporate both skills and knowledge, and that the final output is an operation and not a tangible product.

Precise figures regarding the size of market and individuals involved are notoriously difficult to calculate. This stems from the nature of work being largely carried out by either jobbing type companies, directly employed operatives or for larger facilities, being executed within an overall facilities or serviced building contract. The one thing that can be precisely determined is that the maintenance market is increasing by both size and importance.

According to the BSRIA report *The UK Maintenance Market – Contract Maintenance for Building Engineering Services* the annual growth of the maintenance market between 1995 and 1999 was 12.95%, with the total value of the market rising to an estimated £5.05 billion. The report also highlights the growing importance of term maintenance type contracts. These have been rising by 14% annually to represent approximately 30% of the total market.

Two other interesting statistics can be derived from the report that highlights the growing importance of strategically procuring maintenance works. Of the total market value, £2.5 billion or nearly 50%, is undertaken by direct labour under an in-house type of contract. This demonstrates the importance of the end clients facilities and maintenance managers in being involved with both the initial project briefing and detailed discussions over system and equipment selection.

Maintenance is a problem for the building owners and operators of today, particularly for the majority of buildings that are of average size held by non-building orientated companies. Building size and the nature of the client has a considerable influence on maintenance procurement. In the very large owner/occupier group there is likely to be a resident maintenance management team with a high level of expertise and sophisticated system for maintaining the building. The small building owner would normally have simple services in which even if some part fails completely the building can often continue until a repair is made. The problem area for maintenance is that the vast majority of buildings fall into the middle ground of sophisticated technology held by poorly informed clients.

7.2.2 An Overview of Maintenance Procurement

Unlike capital procurement where the contract is really put in place to cover legal remedies in the event of a dispute, a maintenance contract is a multifaceted document that is used to manage the works. As no tangible product is to be built the maintenance contract must act as the instruction to carry out the work, define the type of work involved and state amount of remuneration to be given.

As the trend for businesses to outsource all non-critical work tasks continues, the complexity of maintenance contracts grows. The growth in both maintenance and facilities management has been directly attributable to large companies wishing to discharge themselves of the responsibility for running non-core functions, while receiving the benefit of assured service level. Although it may be more costly, these same companies benefit from reduced staff levels, while having the ability to call upon the contractors larger skill and resource base in times of need.

Contractors and consultants not only receive the short-term benefits, many have gained substantially as this trend has driven the maturing of the marketplace and the development of maintenance into its own profession.

To be successful, outsource contracts must be specifically written to ensure the desired performance level sought by the facilities owners are achieved.

The overall success of the maintenance strategy will be determined by the skill with which the maintenance contract has been written. As each system will require differing levels and types of maintenance, the contract must reflect this and provide a method to ensure that requirements are explicit. Each system must be maintained to its required performance level. Neglect of some aspects could either lead to elevated operating costs or the premature degradation of key components. This is particularly important with term contracts. Indifferent maintenance could see a system being handed back that has been severely devalued through poor maintenance.

Defining the specific requirements for maintenance must also be considered in the context of legal requirements. The contract must clearly state the minimum level of maintenance required, together with a detailed programme for its undertaking. This is best handled through the use of equipment schedules, with each item of the plant's maintenance requirements explicitly detailed and ideally back-up with reference to an appropriate legislative document. The contract must be written in such a manner to allow competitive, yet appropriate, pricing, while allowing the client a ready means to administer the contract. In short, the contract becomes a multifaceted pricing, administrative and briefing document.

One particular peculiarity of maintenance works is the lack of role for consultants. Although maintenance consultants exist, very few specialise in building services. The vast majority of tenders for maintenance works are drawn up by traditional building services consultants, but a growing number of specialists are beginning to emerge. These specialists are normally ex-client based facilities and maintenance managers who have an in-depth knowledge of maintenance operations.

Like most aspects of construction, the level of service and knowledge held by consultants can be indifferent. Care must be undertaken when selecting such consultants to draft suitable contractual documents and devise both technical and procurement strategies, that they have adequate experience and knowledge of the type of systems involved.

7.2.3 Maintenance Activities

Maintenance is a unique operation as it must combine several specific operations, these range from the strategic to the purely functional. This allows a number of options for the procurement or maintenance, but each party must be jointly linked with a detailed contract.

Typical maintenance activities could include:

- operation of systems or building
- maintaining the system to a specific performance level
- repairs
- component or partial system replacement
- total plant replacement
- call-out services
- record keeping of as-built drawings
- energy monitoring and energy management
- budget planning
- miscellaneous improvement or refurbishment works.

Based on the developed strategy, maintenance can be either procured from a single company on a comprehensive contract, or each system or even each major plant item, could have its own maintenance agreement which is administered through a management type contractor (normally by a facilities management contract). This very much reflects the options that are available for capital construction.

Maintenance can generally be procured by one of four ways, which range from the functional only breakdown attendance, to the strategic management of the overall value of the service's installation through a comprehensive contract:

- **Breakdown service.** This type of maintenance is functional only and represents the lowest level of service available. Repairs are only undertaken when a component fails and no work is undertaken that either prevents failure or improves performance. It best compared to the AA or RAC equivalent for buildings. As the service is neither preventative nor strategic it should rarely be considered for modern buildings, but could be used to supplement an in-house service where access to specialist knowledge and equipment could be useful at a time of complete service failure. An example would be lift or escalator repairs.

- **Preventive maintenance.** Preventive maintenance begins to look at maintenance strategically by anticipating the failure of components and putting forward corrective action prior to their breakdown. The strategy normally follows a set programme of inspections and work tasks that could range from pure lubricating to periodic replacement of components. By these actions breakdowns are supposed to be prevented, while the system is maintained to an acceptable performance standard.

- **Preventive maintenance and breakdown service.** Under this type of agreement, preventive maintenance is extended to provide an emergency service for repairing any breakdowns.

- **Comprehensive maintenance.** Comprehensive contracts are the most commonly used today. Whilst the term differs in meaning , it is generally taken as the provision of all labour, materials, equipment and spares to maintain a particular system, item of plant or complete building to a given set of standards. It represents the most strategic and thorough form of maintenance agreement as the contractor must provide all that is necessary to ensure a system delivers the site performance required by a client. Penalty clauses for poor or under performance are common.

The above types of maintenance arrangements must be considered together with an appropriate contract. As the contract defines both the performance standard and financial remuneration, it must reflect the type of maintenance to be undertaken. Moreover, the general maintenance strategy must be considered at the capital procurement stage to determine how each system will be maintained, as some synergy may be created between the installer and maintainer.

It is also becoming more commonplace to seek a comprehensive installation and maintenance contract, where the installer must maintain the system for an initial period, normally between 2 to 3 years. This must be carefully considered as the installation contractor must be equally competent in both installation and maintenance.

7.2.4 Contract Arrangements

Despite the increasing sophistication of the maintenance market, as yet no standard forms of contract exist for general public use on maintenance contracts. Currents forms of contract are dominated by the GC/Works series published for use by the UK government on their own premises. Therefore most maintenance contracts will be either bespoke or a modified version of a capital installation contract. Great care must be taken within the drafting of such a contract, as no standard construction contract provides a suitable base for maintenance.

As with all contracts, the two primary goals of the contract are to state the level of service to be provided and the amount of remuneration to be made for its delivery. As no tangible product is delivered the contract must explicitly state the level of and manner in which the service is to be undertaken.

- **Facilities Management.** This is the most comprehensive form of contract as it would normally transcend just the maintenance of the system and include the general labour needed to manage the building. This could even extend to such services a caretaker, security or concierge, and may even include the provision of consumables such as toilet paper.

- **Comprehensive.** As outlined earlier, a comprehensive contract normally covers both the preventative maintenance and breakdown service of the system or building. It is important in these types of contract to determine whether the building will be manned fulltime, and if not, what the guaranteed response time will be in the case of failure. It is normal for the contract to include within the stated cost replacement of components to a given value. Although it provides cost certainty to the client, they run the risk of adequate maintenance being carried out as the contractor pays for the components out of his fee. Sufficient checks and performance guarantees must be included within the contract to ensure adequate component replacement.

- **Fully comprehensive.** Similar in fashion to a service level agreement, a fully comprehensive contract is executed by a contractor who, for a fixed fee, undertakes to adequately maintain the system including all component replacement for a set time period.

  A fully comprehensive contract will need an appropriate term to ensure that the premiums taken for full plant replacement risk are likely to be called upon. To have a five-year plant replacement contract would appear to offer more insurance than a business-based plan. Such a contract placed on a 15-year-old building would be a high risk to the contractor. It may also not be possible for the true condition of plant and systems to be determined before the price is agreed, leading to conflict regarding the basis of the contract where the actual requirements for plant replacement significantly exceed what may have been viewed as a reasonable competitive estimate.

- **Continuation.** A continuation contract is one that can be extended beyond its initial period. Normally during the initial set of the contract a costing mechanism, such as the Retail Price Index or the RICS Building Maintenance Index would be used to agree inflation levels. Both sides are able to gain from these types of agreements as continuation of the teams allow some synergy and knowledge of the systems. Further savings are made by the non-necessity of tender documents.

- **Service level agreement.** These types of contracts are unique to maintenance as they do not specify what duties or how they are to be undertaken, but state the level of performance that must be achieved by the system. For example, a service level agreement for an air conditioning/heating system may state that room temperature must be 21°C +/- 2°C. It is the responsibility of the contractor to ensure this is maintained during the stated hours.

  The most complex of service level agreements are those administered under the UK government's private finance initiative (PFI). Major projects are let as a combined construction, finance and service contract for the anticipated lifespan (e.g. 20 years) of a building. The PFI contractor, which is normally a consortium of financiers, contractors and service companies, arrange the finance for the initial cost, carry out the installation and are responsible for its maintenance and operation throughout the full contract term. The client is charged a cost that is based upon the level of service provided within the building. This could be either based upon such things as the number of students attending class in schools, to even the number of operations able to be performed within a hospital.

PFI contracts allocate the risks for design, funding, installation and operation to those best able to manage them, leaving the client (or service user) to concentrate on core business activities. The concept is being applied particularly in the public sector to enable capital developments to take place. It also puts more onerous responsibilities on the development contractor who has to accept the risk of the ongoing viability of the installation over the full life of the contract.

- **Labour only.** Normally an outsource type of contract, a contractor provides only the labour necessary to maintain the system. The client manages and directs the workforce as to the tasks that must be undertaken. The advantages to client being a direct workforce, without the burden of direct employment. Further advantages are gained in that differing types and levels of expertise can be supplied to meet the demands of the maintenance schedule.

- **Inspection and maintenance.** These types of contracts were traditionally known as lube and grease, as the contractor carried out only the lubricating of working parts and the visual inspection to confirm the system was working properly. The contract is normally set up to carry out a set number of visits per year for a flat fee. Any work needed to replace components or undertake repairs were charged additionally to the contract.

- **Planned preventive maintenance.** Planned preventive maintenance is the strategic execution of a maintenance strategy that seeks to minimise the risk of loss of service from the plant and optimise its economic life. The actual strategy could be determined by a number of parties, either the client's own facilities staff, an outside consultant or even the tendering contractor. Each has its merits and detractions. Whilst the use of a contractor or outside consultant can bring advantages in areas of specific expertise or more innovative monitoring procedures, the client still runs the risk of having an adequately maintained system returned at the end of the contract. For installations or clients without the independent knowledge to assess the adequacy of proposals it is recommended that use be made of a standard tasking schedule such as that published by the CIBSE/HVCA Standard Maintenance Specification for Mechanical Services in Buildings.

- **Caretaker maintenance.** This type of maintenance is normally only undertaken on uninhabited buildings that require minimal maintenance on critical systems such as security or fire alarm/suppression systems. All other systems should be drained and properly shut down to minimise health and safety risks from failures and bacterial growth. Obviously the contract could embrace the actually shutting down and eventual start-up of the systems.

- **Call-out only.** Also known as a breakdown contract, the contractor undertakes to provides an agreed response service on the breakdown of a particular system or item of plant. Performance standards are set for response times, with penalty clauses being enacted for achieving them. An all-inclusive labour rate includes for both the call out charge and a hourly rate for repairs. There are numerous ways in which these type of agreements can be structured. Some are based on minimum retainer fees, whilst others are minimum charge out rate with the hourly rate decreasing over time (thereby ensuring the contractor is expedient in the time taken to execute the repairs).

- **Specialist services.** Lifts, escalators, complex automatic controls, major refrigeration plant, fire alarm systems, security systems, uninterruptible power supplies and water treatment are all examples of specialist services within building services maintenance where particular expertise is required. This may be obtained through the main maintenance contractor who arranges subcontracts and incorporates their costs into the charge to the client, or the client may appoint the specialist services contractors directly.

Measured Term Contracts

Traditionally used by government and local authorities, measured term contracts provide for a highly accountable but costly method of executing maintenance. They are generally based on pre-priced schedule of maintenance activities, such as detailed schedule of rates or descriptions of standard maintenance operations, where the contractor states a percentage addition/reduction on the stated rate to undertake the works. Although many authorities use their own schedules, it is common to use one of the publicly available schedules, such as the *PSA National Schedule of Rates* or the RICS's Building Maintenance Index's *Building Maintenance Price Book.*

The advantages of using these is that a number of GC/Works contracts and the JCT Measured Term Contract (1998) are based on using such documents. Furthermore, they are regularly updated with respectively quarterly and monthly indices, allowing a ready form of fluctuation to be entered into a contract.

Most authorities instigate measured term contracts by class of building, which are all encompassing of general building works and services maintenance. These are typified by housing stock maintenance contracts. Larger government and inner city local authorities are more likely to be holders of more commercial type buildings and therefore normally separate their term contracts into specialised operations.

Variations of maintenance term contract can include:

- **Specialist term contract.** Similar in style to a maintenance term contract, a specialist term contract is used to maintain a specific item of plant or specialist system such as lifts, generators and chillers. They can be implemented in a similar manner, but are more likely to be undertaken be either a specialist company or a maintenance division of the original equipment manufacturer.

- **Daywork term contract (DTC).** This type of agreement is only used in situations where a task cannot be identified and scheduled in advance. The contractor is paid on a cost plus basis, where the total hours worked and materials used are costed against an agreed set of daywork rates. These forms of contracts are rarely used as no incentive exists for the contractor to be expedient with the repairs. The advantage to a client is being able to call upon specialist labour to either carry out a specific task or supplement his or her own direct labour force.

- **Planned maintenance term contract.** This form of contract is suitable for the routine servicing of plant and equipment. These contracts typically contain two elements: a lump sum per annum for undertaking a schedule of activities, and a schedule of rates for the rectification of defective components which arise unpredictably during the life of the contract. Under certain conditions "comprehensive" contracts can be placed where both elements are combined with a lump sum and the contractor takes the risk for failures. This is particularly relevant with engineering plant when the servicing contractor can quite legitimately be expected to be responsible for the cost of repairs if not maintained or serviced correctly. Payment for this type of contract is normally monthly in arrears as one twelfth of the annual sum plus computed sums for repairs or after each service visit.

- **Tendered schedule term contracts.** This as an arrangement under which contractors tender against a priced schedule of activities as under a maintenance term contract. This schedule is then used as a basis for measuring in advance the contract price for a maintenance contract. This effectively becomes a lump-sum project contract and it has the advantage of ensuring an element of competition in the price while reducing the overall contractor selection period. Payments are measured, valued and certified as the works proceed in the conventional way. This type of contract would be used primarily when time is short and it can reduce the overall pre-contract stage by running the design and tender stages together, based upon an approximate quantities guide.

7.2.5 Financial Agreements

The contract arrangement determines the manner in which the system is to be maintained. However payment can be by one of several methods. Not all are entirely suitable for maintenance type works, but must be carefully judged against the manner in which the work is to be undertaken, the reasonableness of risk transfer and accountability required by the client. Unreasonable transfer of risk will result in vastly inflated costs and the likelihood of financial retribution by a contractor.

Methods of financial remuneration can include:

- **Lump Sum.** Normally stated as a fixed price figure, the contractor offers to carry out a given set of maintenance operations, including the supply of all necessary labour, components and resources to complete the works. This requires a very detailed schedule of work to be written, but allows the client maximum cost competition. It is rare for maintenance works to be carried out in this manner as few systems can be adequately surveyed to determine the full extent of the works.

- **Estimated Price.** Although still rare, these types of pricing agreements are linked to a risk/reward scheme, where the contractor must complete the works within a given cost framework. Savings within the costs are shared between contractor and client, whilst it is normal for the contractor to take all risk in cost overruns.

  A variation to this type of agreement, known as target cost, is where a schedule of rates is used to first estimate the project, in order to achieve client approval, and the end work is measured again to determine the exact cost.

- **Fluctuating.** As most maintenance contracts take place over an extended timeframe, it is not always appropriate for the contractor to take the risk on possible price fluctuations or inflation. Regardless of the original nature of the price, be it rates or a cost work element, some form of fluctuation can be applied to it. The most common forms in maintenance are the Retail Price Index or the RICS Building Maintenance Index.

- **Cost Plus.** The contractor carries out all work and charges for the actual costs plus a percentage to cover overheads and profit. The danger is that the most cost-effective solution may not always be pursued. It requires very close technical and financial supervision and the cost to rectify faulty work and meet the wages of specialists in short supply, are both problem areas. This payment method can be of the form where the client sets the base rate and the tenderers' offers are purely their individual percentage mark up/down.

- **Bill of Quantity.** Although more common for capital construction works, their suitability for maintenance should be questioned. Bills of quantities rely on the breakdown of each work task into separate elements based on the trades and nature of work. Multi-skilled operatives, usually in environments that are confined and difficult to describe normally carry out maintenance tasks. Therefore, the pricing of work on an elemental basis would be difficult. Furthermore, the Standard Method of Measurements normally based on construction works, again not reflecting the true nature of maintenance. What bills of quantities do offer though, is a detailed schedule of labour and materials rates to complete a given set of work. By using singular elements a composite price can be built up for most types of work.

- **Schedule of Rates.** Rather than bill of quantities, the more favoured document to value work is a schedule of rates. Based on a composite description of work, the contractor prices for and is paid to undertake the complete works of a particular maintenance operation. Unlike bills of quantities, the price stated allows for all necessary operations, including such tasks as painting and insulating.

A variation on this would be where the client supplies a priced schedule of rates and contractors bid to carry out the works on plus/minus percentage rates. Use can be made of pre-written schedule, such as those published by the *PSA National Schedule of Rates* or the RICS's Building Maintenance Index's *Building Maintenance Price Book.*

The type of the contract selected should match the clients needs for service levels and cost expectations. The cost of administration of measured term, specialist term and dayworks can be high. The cost particularly depends on the extent to which the employer chooses to check the measurement of the works undertaken by contractors.

A sampling process is commonly used comprising a proportion of between 10–20% of orders raised. If such sampling shows a disproportionate rate of error then the sampling percentage should be increased. Sampling should be on a random basis with no prior indication given to the contractor.

A comprehensive contract is an extension of the preventive contract with the service provider also agreeing to repair or replace any piece of plant or equipment that breaks down during the contract period. The lump sum contract price generally covers repairs up to a certain value and there is no administrative cost for raising orders for such work. The lump sum price gives the client greater control over the maintenance budget, with a known outlay of expenditure. Over the contract period the risk element of major plant and equipment breakdowns is borne by the service provider.

7.2.6 Standard Forms of Maintenance Contracts

Traditionally, contracts that were specific to maintenance were only available through the UK government's Property Services Agency (PSA). As the maintenance market matured and private clients sought to outsource their maintenance works, they developed their own contracts, as no commercial contracts were available. In more recent times, the Joint Contracts Tribunal (JCT) published two forms of contract, but neither are specific to building services.

Currently only the new forms of government contracts, and particularly the GC/Works/9 series covers building services specially, although the others may be suitable, but are dependent upon the exact nature of the works.

Along with the normal constituents of any contract, being definitions, method payment, services to be provided, date of commencement and duration and how the contract will be terminated, maintenance contracts must adequately schedule out the activities to be undertaken. Careful consideration must be given to this scheduling in order to address the following issues:

- the performance of both the contractor and facilities will need to be measured and monitored;
- responsibility for determining when plant replacement is due, together with who specifies the replacement's performance will need to be stated;
- the terminology will need to be clearly define to ensure no misunderstanding;
- risk transfer of both system performance and component failure will need to be explicit, as will any payment for damages caused by failure.

A new set of government forms of contract that can be used for the operation and maintenance of building services plant has recently been introduced by PACE, the Property Advisors to the Civil Estate. These include:

- **GC/Works/6 (1999) - Daywork Term Contract.** This is a term contract intended for procuring work of a jobbing nature. The labour charge in GC/Works/6 is based on hourly rate rather than on the wages paid in accordance with the wage-fixing bodies. Materials are paid for on the basis of cost plus a percentage addition. The publication containing the conditions of contract also incorporates model forms and a commentary.

- **GC/Works/7 (1999) - Measured Term Contract.** It is a standard form of measured term contract based on a schedule of rates, which is provided by the employer. Orders are placed with the contractor as and when required. The typical contract period is three to five years. The publication containing the conditions of contract also incorporate model forms and a commentary.

- **GC/Works/8 (1999) - Specialist Term Contract for Maintenance of Equipment.** It is a specialist term contract for use where specified maintenance of equipment is required and can be costed per task. One of the principal differences between this form and GC/Works/9 (see below) is that the price is not based on a single annual cost. Under this form the contractor prices a schedule of work and interim payments are based on the measured work actually carried out. The typical contract period is expected to be three years. Model forms and a commentary are also provided along with the conditions of contract.

- **GC/Works/9 (1999) - Lump Sum Term Contract for Operation, Maintenance and Repair of Mechanical and Electrical Plant, Equipment and Installations.** It is a lump sum maintenance term contract for the operation, maintenance and repair of fixed mechanical and electrical (M&E) plant, equipment and installations. This form of contract may be used in a single establishment or a complex of buildings close enough to be conveniently covered by a single contract. The term of the contract is for a period of one to five years. The contract price includes one-off repairs up to a specified maximum cost per repair. Model forms and a commentary are also provided along with the conditions of contract.

There are other forms of contract in the industry such as the JCT forms of contract that can be used for maintenance purposes. Neither of the two forms available are specific for building services and are primarily written for general building works. The two most suitable forms are:

- **JA/C 90 Conditions of Contract for Building Work of a Jobbing Character Nature:** This is a short form of contract for use by experienced clients to order works of a simple nature. The contract is suitable for use by both private and government clients. As the conditions of contract are short and simplified, they are not suitable for projects that require either long-term maintenance or will be undertaken over extended time periods. Primarily it would only be suitable to place an order with a contractor to cover interim plant replacement or work of a singular one-off type.
- **MTC 98 Standard Form of Measured Term Contract:** Apart form the GC/Works series, this form of contract provides the most comprehensive contract suitable for maintenance-type works. Its conditions of contract and schedules are suitable for minor works and interim plant replacement. Although written for general building works, minor amendments would make it suitable for building services maintenance.

The only other commercially available form of contract that may be suitable for building services is the Chartered Institutes of Building's Facilities Management Contract 1999. Although primarily written for the actual management of a complete building, together with its related service supply, the contract is written from the context of a service provision to maintain a function. Where its suitability does diminish is in the detailed explanation and pricing of breakdowns and planned maintenance.

The author is indebted to the assistance of Rohan Nanayakkara in the preparation of this chapter.

REFERENCES

1. Samuelsson-Brown, G. and Parry, T. (2000) *The UK Maintenance Market – Contract Maintenance for Building Engineering Services*, Bracknell, BSRIA Publications.
2. CIBSE (2000) *Guide to Ownership, Operation and Maintenance of Building Services*, London, CIBSE.
3. DETR (1987) *C910 Lump Sum (M&E) Operation, Repair and Maintenance Contracts - Guidance Note for the Preparation of Particular Specifications*, London, Property Services Agency.
4. Economic Commission for Europe (1987) *Guide on Drawing Up International Contracts for Services Relating to Maintenance, Repair and Operation of Industrial and Other Works*, Brussels, United Nations Publications.
5. Hanson, M. (1999) *CMS: Guide to Facilities Management Contracts*, London Asset Information Ltd.

6. HMSO (1990) *General Conditions of Contract for Building & Civil Engineering*, London, HMSO.
7. Horgan, M.O. and Roulston, F.R. (1977) *The Elements of Engineering Contracts*, London, W S Atkins & Partners.
8. PACE (1999) *GC/Works/6 Daywork Term Contract*, London, HMSO.
9. PACE (1999) *GC/Works/7 Measured Term Contract*, London, HMSO.
10. PACE (1999) *GC/Works/8 Specialist Term Contract for Maintenance of Equipment*, London, HMSO.
11. PACE (1999) *GC/WORKS/9 - Lump Sum Term Contract for Operation, Maintenance and Repair of Mechanical and Electrical Plant Equipment and Installations*, London, HMSO.
12. PACE (1999) *GC/Works/10 Facilities Management*, London, HMSO.
13. RICS (2000) *Building Maintenance: Strategy, Planning & Procurement - Guidance Note*, London, RICS Books.
14. Smith, M.H. (1992) *Maintenance Contracts for Building Engineering Services – a guide to management and documentation. Application Guide 4/89.2*, Bracknell, BSRIA Publications.

Chapter Eight

Product Modalities

8.1 PRODUCT MODALITIES

The product, being the culmination of the required systems and physical installations, is the purpose of undertaking the procurement exercise. Although the vast majority of the product will be defined once the project team is on board, sufficient development must be undertaken to ensure the parties forming the team have the appropriate abilities and expertise.

The key considerations that are presented here will affect procurement as particular organisations with specific abilities will need to be recruited.

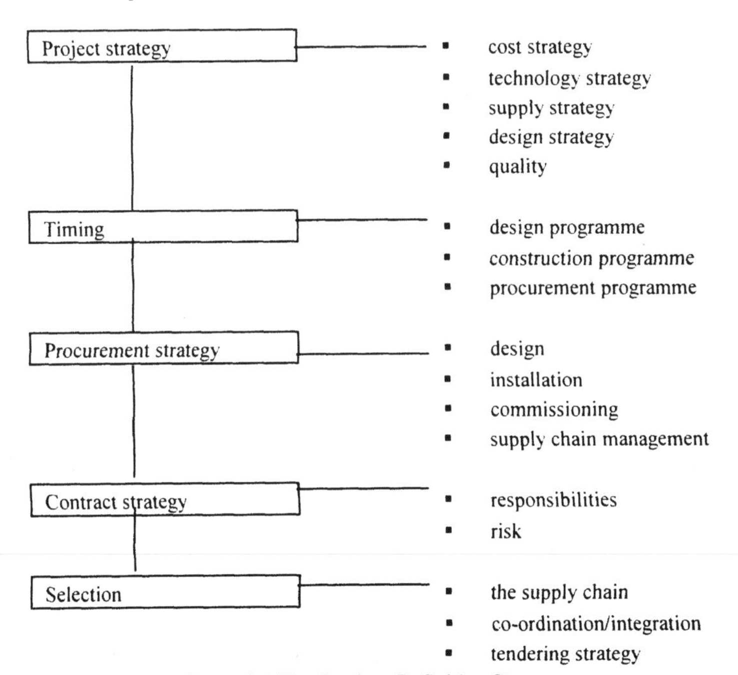

Figure 8.1 The Product Definition Stage

8.2 MODULARISATION AND PREASSEMBLY

The decison to use modularisation or preassembly must be a balance between the additional costs and co-ordination required to successsfully implement it and the added benefit of quicker construction and predetermined quality. These considerations can be complex and can best be seen using the the diagram below.

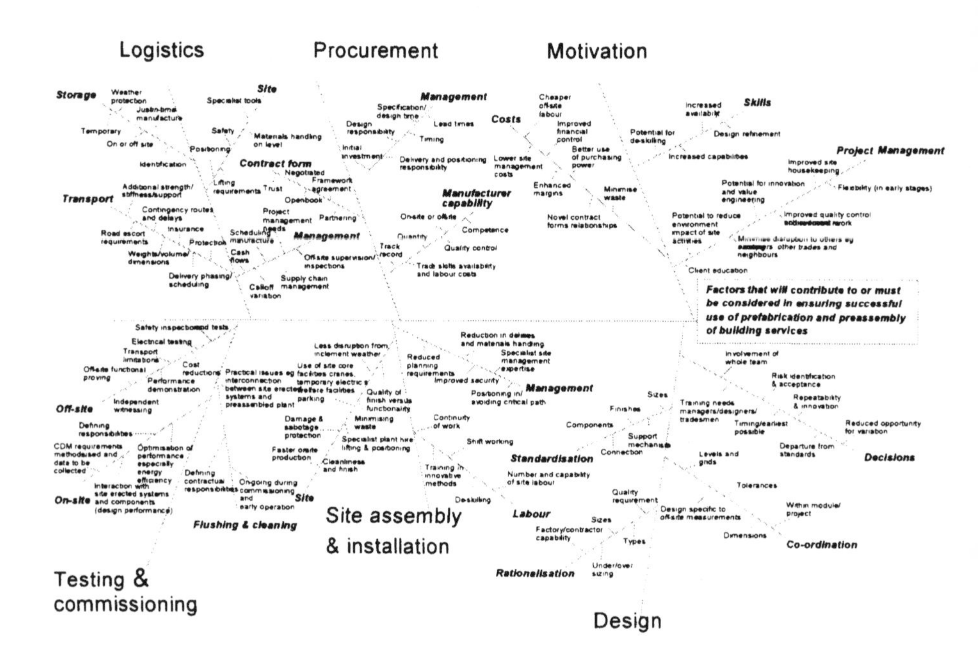

Source: Prefabrication and Preassembly ACT 1/99, BSRIA Publications

Figure 8.2 Factors Affecting Preassembly

Each of these factors contributes to the decision, which is one that can be viewed as a pure business judgement. To ensure the decision is unbiased, a weighted evaluation technique can reduce the subjective criteria to an analytical level where the subjective can be judged equally with the technical and monitory criteria. The result is a detailed score that can be used to either decide between modularisation and conventional construction or between two differing levels of modularisation.

A pragmatic decision must be made to use modularisation and preassembly, based on the desirability for its use, the project objectives and the likelihood of achieving them. A pro-forma decision matrix has been developed that assesses all of these issues and converts them into a success score.

	Modularisation Consideration	Objective Factor	Likelihood Factor	Project Score
1	Time advantage required	0.9	70	63
2	Limited installation window available	0.2	65	13
3	Strict quality control required	0.8	55	44
4	Limited site storage	0.3	30	9
5	Adequate lead-in design times	0.1	60	6
6	Scope for standardisation	1.0	50	50
7	Labour availability	0.5	40	20
8	Level of repetition	0.3	50	15
9	Advantage in site work minimisation	0.7	85	60
10	On-site fabrication space available	0.4	30	12
11	Congested site working area	0.4	50	20
12	Flexibility in specifications	0.8	80	64
13	Possibility of combing services support with other project requirement	1.0	70	70
14	Are other building elements modular?	0.3	60	18
15	Possibility of future services upgrading requiring modular replacement of plant	0.5	50	25
16	Waste minimisation desirable	0.8	40	32
17	Safety hazards reduced through limited site working	0.6	80	48
18	Buildings services team experience using modularisation	0.6	75	45
19	Availability of suitable modularisation contractor	0.8	90	72
20	Access for prefabricated modules	0.7	70	49
	Total score	11.7	Total score	735
	Perfect score	1170	Rating	63

Perfect Score = 100 x total desirability factor
Rating = project score ÷ perfect score x 100
Modularisation Consideration = variable depending upon project and location
Objective Factor = scored between 0.1 and 1.0 based on project objectives
Likelihood Factor = likelihood of achieving consideration
Project Score = objective factor x likelihood factor

Rating Interpretation
0-20 limited likelihood of achieving benefits from modularisation
21-35 moderate benefits available
36-65 wide range of benefits available
66-80 potential for substantial benefits
81-100 perfect project for modularisation

Source: Prefabrication and Preassembly ACT 1/99, BSRIA Publications

Figure 8.3 Decision Matrix

When considering individual systems, manufacturing production indices could be used:

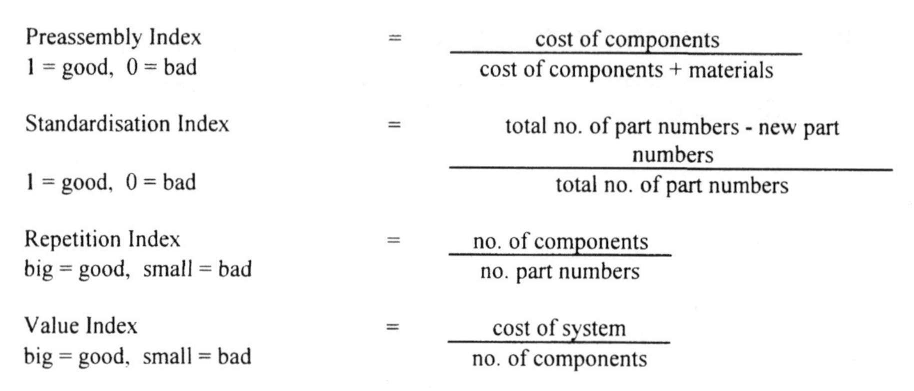

Preassembly Index
1 = good, 0 = bad

$$\text{Preassembly Index} = \frac{\text{cost of components}}{\text{cost of components + materials}}$$

Standardisation Index
1 = good, 0 = bad

$$\text{Standardisation Index} = \frac{\text{total no. of part numbers - new part numbers}}{\text{total no. of part numbers}}$$

Repetition Index
big = good, small = bad

$$\text{Repetition Index} = \frac{\text{no. of components}}{\text{no. part numbers}}$$

Value Index
big = good, small = bad

$$\text{Value Index} = \frac{\text{cost of system}}{\text{no. of components}}$$

Figure 8.4 Manufacturing Production Indices

8.3 WHOLE LIFE COSTING

Whole life costing is a complex issue for building services as it must combine the technology and risk assessments used to develop a maintenance regime with the complexity of economics. What follows is a simplified guide to the process, but prior to undertaking such studies reference should be made to publications listed under References.

Two objectives should form the basis for undertaking a whole life costing exercise at the procurement stage of a project. Firstly, it should determine the emphasis that the client will place on the balance between whole life and capital cost, both within product selection and over system design. Secondly, it forms the basis for determining the balance between cost and subjective requirements.

Ideally, the assessment should follow a six step process. This process assumes that a number of alternative design and procurement options are being considered. If a single option is already decided, then the criteria weighting and assessment can be ignored. The six steps are as follows:

Step 1: Assembly of Relevant Information and Decision Criteria

The information collected should include all that necessary to make an objective assessment of the costs and issues involved. Principally this will include information on:

- The description of the design option being considered;
- The expected life of the building, the system and individual components;
- The discount rate to be applied. This could include the loan rate, expected level of inflation and investment rate;
- The maintenance strategy to be employed, including replacement cycles of components;
- The estimation of total capital cost, split down between each system.

The expected life will vary depending on what it is. For example, the overall life cycle of building is likely to be between 25 to 100 years, where as for any building services system, it is uneconomical and technically improbable that a system can be designed for a life of greater than 25 years. Furthermore, within each system will be components with life cycles of anywhere between 3 months (say for filters) to 10 years (say for pump motors).

This process of balancing the differing life cycles is known as service life planing. Officially it is known as the design process which seeks to ensure, as far as possible, that the service life of a building will equal or exceed its design life, while taking into account (and preferably optimizing) the whole life costs of the building. The standard to which life cycles should be designed is listed under ISO 15686, which provides a methodology for forecasting the service life and estimating the timing of necessary maintenance and replacement of components. It thereby provides a means of comparing different options.

For individual components, the methodology outlined in ISO 15686 can be used or reference should be made to a standardised list, such as that printed by the CIBSE in their publication *Guide to Ownership, Operation and Maintenance of Building Services.*

Although it is recognised that many buildings and systems will have a useful life span over more than 25 years, studies should limit themselves to this time frame due to the uncertainties and escalating cost levels beyond this period.

Typical cost elements to be considered in a building services whole life cost exercise are shown in Table 8.1.

Capital Costs	
▪ services installation	▪ design team fees
▪ builder's works	▪ statutory consents
▪ site set up (accommodation, storage etc.)	▪ health and safety
▪ taxes (vat etc.)	▪ plant and equipment
▪ other services (supply authority contributions)	▪ demolition and site preparation
Finance Costs	
▪ finance future maintenance	▪ finance for construction costs
Operational Costs	
▪ fuel	▪ attendance
▪ insurance	▪ storage
Maintenance & Repair Costs	
▪ preventive maintenance	▪ corrective maintenance
▪ replacement	
Residual Values	
▪ salvage value	▪ disposal fees and charges

Table 8.1 Building Services Cost Elements

The final pieces of information are the discount rate, rate of inflation and escalation rate. Normally the client will provide all of these, as it is their money that is being used.

As a rule of thumb the following should be used:

- **Discount rate:** the rate is equal to the financing rate for the project, or 1-2% above the Bank Base Rate or the annual rate of return a business expects to receive in gross profit. For government projects a standardised rate is normally set by the Treasury, which at the time of writing is 6%.
- **Rate of inflation:** trying to determine what the rate of inflation is over 25 years borders on pure guess work.
- **Escalation rate:** the escalation rate is the rate in which a sum of money must be invested today to meet an expenditure in the future. It differs from discount rate as the objective is to achieve a lump sum in the future and therefore is largely affected by the lump sum initially invested.

Step 2: Determine Design Option

An objective assessment must be made as to the best possible option to meet the design requirements. This will normally be comparing the minimum acceptable design option against the others.

Standard criteria, which are common requirements for any project, could be used for this assessment. These may include the following:

- durability expressed as probable service life
- ease of repair
- ease of replacement
- energy efficiency
- environmental issues
- flexibility (adaptability)
- lead-time to purchase
- low maintenance
- quality
- reliability.

In addition to these standard criteria an additional set specific to the project can also be used. Each criteria should have an appropriate technical measurement for comparison and demonstration that the set criteria will be achieved.

Step 3: Set Assumption and Decision Criteria

Assumptions will include the criteria set out within maintenance strategy that will affect the whole life cost, together with the various rates decided on in Step 1. Maintenance assumptions applicable to whole life costs will include:

- the overall maintenance strategy for each system
- the use of in-house or external services company to provide the maintenance
- the acceptable level of failure risk for each critical component

- determination of the point of failure for each component
- likely timescale for outlet replacement, based on refurbishment schedule
- economic threshold for performance upgrade of a component or system.

For each criterion established in the previous step, a weighting which reflects its relative importance out of a total of 100 must be established. For example, if durability is very important it may attract a rating of 20 out of 100. Flexibility/adaptability may be less important, with a weighting of only 5. The total scores allocated across all criteria should add up to 100. Therefore, for a particular electrical system the criteria may score, as in Table 8.2 below.

Criterion	Scoring
▪ Durability	20
▪ Ease of repair	10
▪ Ease of replacement	5
▪ Energy efficiency	15
▪ Environmental issues	5
▪ Flexibility	5
▪ Lead-time to purchase	0
▪ Low maintenance	10
▪ Quality	15
▪ Reliability	15
total	100

Table 8.2 Criteria Scoring

Step 4: Options Scoring

Each design option should be scored out of 10 for each criterion. If a particular aspect of the specification has been previously identified as essential and the option has failed to meet it, the whole option will fail automatically. The criteria above and for four further theoretical options may score as shown in Table 8.3:

Criterion	Scoring	Option 1	Option 2	Option 3	Option 4
▪ Durability	20	20	15	15	10
▪ Ease of repair	10	5	10	8	8
▪ Ease of replacement	5	3	2	2	2
▪ Energy efficiency	15	14	x	13	10
▪ Environmental issues	5	5	5	2	2
▪ Flexibility	5	0	5	4	3
▪ Lead-time to purchase	0	0	0	0	0
▪ Low maintenance	10	5	10	8	8
▪ Quality	15	12	12	10	x
▪ Reliability	15	10	5	7	8
total	100	74	X	69	X

Table 8.3 Criteria Scoring of all Options

Analysis shows that Options 2 and 4 are excluded as they fail on two criteria.

Step 5: Identify Cost Profile

From the cost established in Step 1, each will occur at different times and frequencies throughout the life of the system. Step 5 seeks to regularise all of these costs through a standardisation of time. This is accrued through a discounted cash flow.

The basic formula used to calculate the present value £X in t-years time with a rate of interest of r%.

$NPV = X/(1+r)^t$ where:

NPV = net present value
X = initial capital sum
r = rate of interest
t = period of time

This formula can then be rearranged to calculate the future value £X in t-years time with a rate of interest of r%.

$FV = Xx(1+r)^t$ where:

FV = future value
X = initial capital sum
r = rate of interest
t = period of time

Further rearranging can provide the formula for calculating the present value of and annual expenditure £X over t years at an interest rate of r%. An escalation rate can be used if it is decided that price rises will be above inflation (this is normal practice for such costs as fuel).

$PVA = Xx(1-(1+r)^t)/r$ where:

PVA = net present value of annual exp.
X = yearly sum
r = rate of interest (escalation rate)
t = period of time

Each cost identified must be reduced to the same present using the above formulas. The escalation of cost against conflicting percentages, such as when inflation and investment interest are both considered cause a myriad of formulas to be developed from the above. Reference should be made to text highlighted at the end of the chapter for a more detailed understanding of discount economics.

Step 6: Select the Preferred Option

The summary must draw together the information and results from the previous steps. For each option the summary must contain:

- A score against the qualitative criteria
- The whole life cost of the item or project
- Identification of any option which has failed to meet any essential requirement.

Based on the stated project objectives, the best option meeting both quality and price requirements should be selected.

As with any procurement decision the lowest cost option will not necessarily be the most appropriate. For example, if an option fails to deliver an essential part of the technical specification then it will not meet users' requirements, and should be rejected even if it does turn out to be the cheapest option.

Worked Example

This worked example is based on a small office building that is currently designed using a variable air volume (VAV) system with fan coil units. It is obviously a considerably simplified version of the calculations and decisions that must be made.

Although in whole life costing it is normal to discount all costs to the present day, sometimes it useful to see the total committed cost. For energy and maintenance, escalated costs allow for ready comparison against service level agreements and highlights to the client the total commitment to ownership. This example is based on a total commitment of ownership calculation, which shows the whole cost of owning and financing the installation over a 25-year period. For simplification, financing costs are ignored.

Three options can be evaluated as to which provide for the best whole life solution.

Options	**1**	**2**	**3**
Proposed System:	VAV and chilled beams	Constant air volume	VAV and fan coils
Replacement Cycles: (simplified consideration of major plant)			
	16 years chilled beams	8 years air handling unit	10 years fan coils
Project Information: (same for all options)	discount rate: 8% rate of inflation: 3% escalation rate: 6%		
Capital Cost:			
system	946,000	647,000	732,000
electrical	120,000	100,000	160,000
total	1,066,000	747,000	892,000
Interim Replacement Cost:			
	145,000 (year 16)	62,000 (year 8 and 16)	186,000 (year 10 and 20)

These costs must be adjusted to future value (FV) for a like-for-like comparison. This is accomplished using the formula:

$NPV = X*(1+r)^t$ where: NPV = net present value
X = initial capital sum
r = rate of interest
t = period of time

For Option 1 this equals:

$NPV = X*(1+r)^t$ where: NPV = net present value
$NPV = 145{,}000/(1+.03)^9$ 145,000 = initial capital sum
$NPV = 189{,}192$ 3% = rate of interest
9 = period of time (25-16)

Discounted Interim Replacement Cost: (based on NPV calculations above)

189,192 (year 16)	78,539 (year 8)	249,968(year 10)
	99,492 (year 16)	335,936 (year 20)

Annual Maintenance Cost:

20,000	16,000	18,000

Annual Fuel Costs:

24,000	36,000	28,000

This calculation can be accomplished through:

$PVA = X*(1-(1+r)^t)/r$ where: PVA = net present value of annual exp.
X = initial capital sum
r = rate of interest (escalation rate)
t = period of time

For Option 1, this equals:
$PVA = 20{,}000*(1-(1+.0\ 6)^{25})/r$ PVA = net present value of annual exp.
PVA = 20,000 = yearly sum
6 = rate of interest (escalation rate)
25 = period of time

Escalated Annual Maintenance Cost:

1,097,290	877,832	987,561

Escalated Annual Fuel Costs:

1,316,720	1,975,080	1,536,173

Total Commitment to Ownership:

Option	**1**	**2**	**3**
Capital Cost:	1,066,000	747,000	892,000
Interim Replacement:	189,192	78,539	249,968
		99,492	335,936
Maintenance:	1,097,290	877,832	987,561
Fuel:	1,316,720	1,975,080	1,536,173
Total	**3,669,202**	**3,777,943**	**4,001,638**

From the detailed analysis, it can be seen that Option 1, although being the most expensive capital cost provides the lowest whole life cost. It has the further added benefit of only requiring one interim replacement of major plant.

For a true comparison to be made builders work and the cost of finance would have to be added.

8.4 TOTAL QUALITY MANAGEMENT

As any construction project, other than a serial project, will involve the unique assembly of people, it can be viewed as a distinct business organisation. To achieve quality harmony must exist between people, processes and the responsibilities they hold. Organisations will view each differently and therefore a consensus of opinion must be established first to form an agreed medium.

Ideally, agreement should follow a structured programme. Typically this would involve:

Step 1: Gain Commitment from all Project Team Partners
Although lead by the client, all members of the building services team must agree that a form of continuous improvement is desirable on the project and each is willing to work towards it. Commitment must focus on the fact that all projects, despite being typically well managed, can be improved. A team workshop and "getting to know you" days will often flush out past indifferences and establish the root cause of typical problems.

Step 2: Develop a Shared Vision for the Project
Construction projects benefit from the fact that they have a well defined tangible mission - they must be built and function properly. All actions taken and processes used must contribute to achieving this mission. In keeping with this mission a mission statement must be developed that highlights the need for change and how it can be brought about within the project.

Typically questions that can be answered within the mission statement include:

- Does it contain the need that must be fulfilled?
- Is the need worthwhile in terms of commitment resources?
- Is it sufficiently long term to create a definitive step change in improvement?
- Is the entire project able to commit to it?
- Does the purpose remain constant through the duration of the project?
- How can each company benefit from its achievement?

Step 3: Define the Measures against which Improvement will be Judged
Improvements must be tangible and the project must have a defined level of improvement. Reference should be made to the KPI's and benchmarks in Chapter 11.

Step 4: Develop the Mission into Critical Success Factors

To achieve any mission, supporting actions must be executed properly . These are known as critical success factors. Typical critical success factors for a building services installation may include:

- having the right-first-time attitude
- having the correct resources
- having motivated, skilled people
- having the correct information in a timely manner
- having zero defects.

Step 5: Breakdown each Critical Success Factor into Key Critical Processes

This is the critical stage in total quality management as changes in a process are the only means in which improvements can be made. Similar to function analysis diagrams used in value engineering, outlining the processes required to achieve a critical success factor requires the listing of the steps and the use of verbs to describe the necessary action.

Step 6: Monitor and Adjust the Processes to Align with Change

The project organisation must become a learning organisation that is willing to learn by change and understand its benefits. This will require constant revisiting of process descriptions and adjusting them to suit the lesson learned.

8.5 VALUE ENGINEERING

Value engineering is a useful tool during procurement to ensure the objectives and requirements set down by the client are robust in their underpinning thought. It can be used to assess the suitability of the overall project option or the specific systems used within the building services.

Value engineering is commonly known for its 40-hour workshops. While appropriate for major projects, a number of other techniques can be employed, for either entire projects or specific elements. Either one or a combination of the four can be employed on any project.

8.5.1 Value Engineering Studies

The Charette

A charette is a single short workshop that seeks to rationalise a client's brief by identifying the functions of the key elements and spaces identified within the project brief. The principle is based on the theory that briefs state an amalgamation of wants and needs, rather than true requirements. A charette attempts to identify and separate. Charettes are carried out along job plan lines:

- first stage being information gathering from original sources rather than accepting what is stated;
- second stage is to be creative with work space arrangements - ideas are recorded with the best incorporated into the brief.

The advantages of a charette include:
- inexpensive
- considered as best way of briefing whole design team
- carried out early in process where major cost influences are decided
- carried out in short period of time
- cuts out and crosses organisational, political and professional boundaries.

40-Hour Study

A facilitated 40-hour study is the most widely accepted approach. It involves a review of sketch design by a second team of independent experts.

Stages of Study

Monday: phase 1 information
- sketch designs and cost estimate are proposed to team
- facilitator states objectives for study and original designer and client present proposals to team
- team then identify main functions of building

Tuesday morning: phase 2 creativity
- brainstorm session to satisfy functions

Tuesday afternoon: phase 3 judgement
- team decides which ideas are worthy of development
- architect may be invited back to review possible proposals

Wednesday and Thursday: phase 4 development
- small groups are formed to develop ideas into detailed proposals

Friday: phase 5 recommendation
- proposed solutions are reviewed for feasibility with short list drawn up for best ideas.

This form of value engineering is considered effective because:
- alternative solutions are costed in both initial and whole life terms
- date is fixed for sketch design
- cost of study is small compared to benefits gained
- failure rate is extremely low - 2% normally
- original design must be judged in order to gauge its effectiveness
- study can take up to 4 weeks of the design time.

Value Engineering Audit

This type of audit is carried out within a project team or company to review major elements of cost expenditure. An independent expert is brought in over a one- or two-day study to determine effectiveness of money being spent. Essentially the audit is an independent justification of a business case for expenditure.

8.5.2 Value Engineering Methodologies

Each methodology can be used at a particular stage of the design to ensure the project is in line with the client value hierarchy or in conjunction with the studies listed in the previous section (see Figures 8.5 and 8.6).

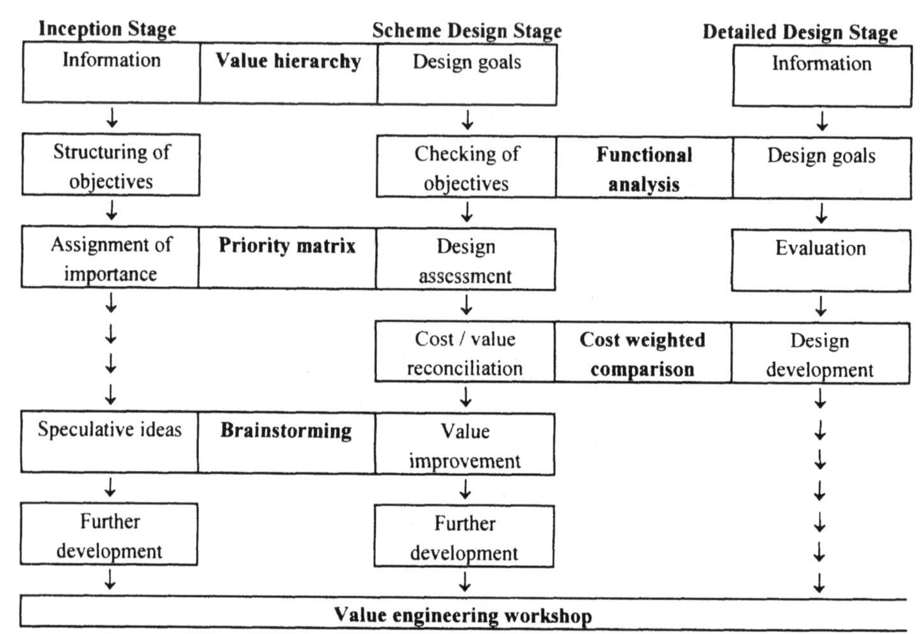

Figure 8.5 Design Stage and Methodology Use

Primary Objective	Secondary Objective	Third Order Objectives		
User comfort	Attractive lighting	Controllable	Varied lighting level	Colour rendering
	Suitable warmth	Controllable	User defined	
	Suitable coolness	Controllable	User defined	
	Minimal stuffiness	Controllable	Moisture	Maximise fresh air
	Minimise construction	Simple systems	Minimum distribution	
	Minimise running cost	Low energy use	Flexible	Future proof

Figure 8.6 Value Hierarchy

Procedure for constructing a value hierarchy.

- Primary objective is positioned to the far left. This must be the clients ultimate mission (see Chapter 4 for strategy hierarchy), and for building services it would normally be based on occupant comfort.
- The secondary objective must be the key elements that must be fulfilled to achieve the primary objective.
- Third order objectives are then derived from the secondary objective and once again are the key elements that must be fulfilled to achieve the secondary objective.
- If felt necessary the tree can be extended to create fourth and subsequent objectives.

Priority Matrix

Each objective identified in the value hierarchy above will not have equal importance to the project success. To understand an objective's importance and the amount it may be compromised by a more important objective, a level of priorities needs to be established.

Secondary and third order objectives are compared separately on a matrix that compares each objective against the other. The level of importance is determined by a 1-4 number system representing:

1 = equal or not as important
2 = slightly more important
3 = moderately more important
4 = considerably more important.

The total score is the sum of that objective's rating against the others. The weighting is then derived from the objective's total score as a percentage of the overall total score for all objectives.

		Matrix Scoring						**Total Score**	**Weighting**
Secondary Objective		**a**	**b**	**c**	**d**	**e**	**f**		
Attractive lighting	**a**	-	1	1	1	1	1	5	10%
Suitable warmth	**b**	4	-	2	3	1	1	11	20%
Suitable coolness	**c**	3	1	-	3	3	3	13	24%
Minimal stuffiness	**d**	2	1	1	-	2	3	9	16%
Minimise construction	**e**	3	1	1	1	-	2	8	15%
Minimise running cost	**f**	4	1	1	2	1	-	9	16%
							Total	55	100%

Figure 8.7 Priority Matrix

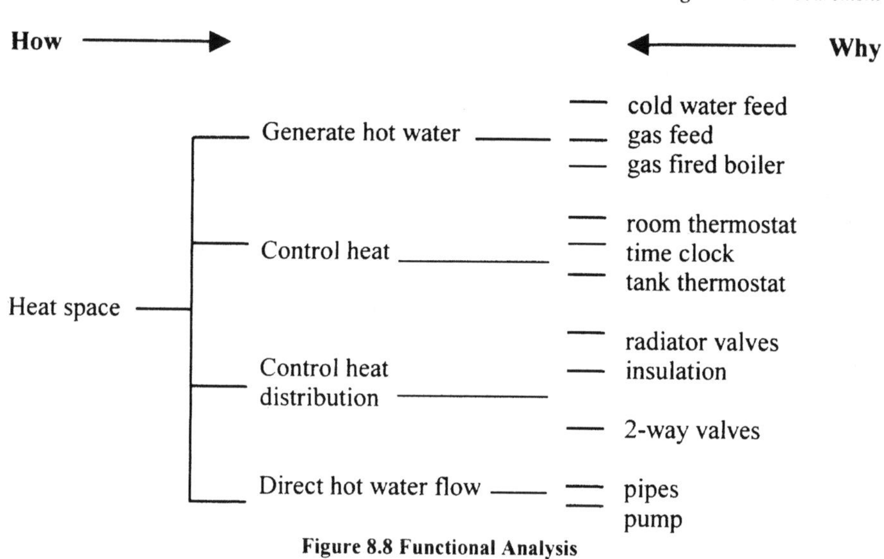

Figure 8.8 Functional Analysis

A functional analysis is constructed to determine that the design solution developed for a particular objective meets the criteria without extravagant features. The analysis is constructed so when read left to right is shows the how the objective is to be determined and when read right to left it answers the why .

REFERENCES

1. Flanagan, R. and Norman, G. (1983) *Life Cycle Costing for Construction*, London, RICS Publications.
2. Hayden, G.W. and Parsloe, C. (1996) *Value Engineering in Building Services*, 15/96 Bracknell, BSRIA Publications.
3. Kelly, J. and Male, S. (1993) *Value Management in Design and Construction*, London, E&FN Spon.
4. Oakland, J. (1995) *Total Quality Management*, Oxford, Butterworth-Heinemann Limited.
5. Wilson, D.G., Smith, M.H. and Deal, J. (1999) *Prefabrication and Preassembly, ACT 1/99*, Bracknell, BSRIA Publications.

FURTHER READING

For further information and detailed explanation of discount economics consult:
Lumby, S. (1994) *Investment Appraisal and Financial Decisions*, London, Chapman & Hall.

Chapter Nine

The Supplying Market

9.1 THE MARKETPLACE

The construction market is one of buoyancy and often fluctuates with the general economy. As building services is a proportional element of the entire construction market, its general fortunes follow the market. With the increasing sophistication of procurement strategies by clients, building services have now begun a transformation where they have evolved into their own industry, separate from the general economies of construction. In the main this is due to the increasing use of service-backed maintenance and facilities management, providing most members of the services team with a proportional income free from the volatility of pure construction.

Generally the building services market in the UK is valued at £18 billion and represents 20% of the total value of the construction industry. This is split between the three markets of general building, housing and infrastructure, and the three project types of new construction, improvement works, and repair and maintenance. This diversity of both markets and work type provides the industry with some safeguard against excess work level fluctuations.

Nevertheless, the general downturns in both work level and profitability since the late 1980s have begun to have serious consequences. There is now a serious shortage of skilled tradespeople and entrants into apprenticeship schemes.

Similar conditions now exist within the consultancy market. A general decline in students studying for degrees in the sciences and engineering has resulted in a shortage of qualified engineers. This, combined with the eradication of fee scales, has produced considerable competition, resulting in lowered profits and lack of inward investment.

The largest change that has occurred in recent years has been that concerning main contractors. As the market for their work has matured, a number of leading firms have either merged or exited the market. This has mainly been due to a mature market where increased risk is being placed upon the contractor, with declining profitability. In an attempt to control this, a number of contractors have entered the building services market.

Their entry has not been into main stream contracting, i.e. executing the work, but has introduced a more appreciative understanding of building services. Direct procurement of plant, together with increased staffing of engineers, commissioning managers and general management, has allowed main contractors to undertake the role of the principal building services contractor.

At the same time the general building services industry has seen increased specialisation driven by both developments in technology and increased skills shortage.

All of this has affected the procurement of building services. Main contractors now possess greater expertise, allowing them to procure building services in a more diversified fashion. They are able to use work breakdown structures to procure specialists directly, thereby allowing their clients a greater choice of procurement options.

9.2 THE ACTORS AND THEIR ROLES

9.2.1 The Client Team

Determining the Client

Determining exactly who the client is, is not an easy process. Depending upon the contractual relations and procurement strategies, the client could be any number of people or organisations, for the term "client" can refer to a number of service receivers.

Traditionally, the person who contracted with the main contractor and hired the professional team was considered to be the client. With simple domestic sub-contract arrangements, the subcontractor believed the main contractor was their client. This followed the general management principles of TQM, which state the next person in the supply chain is the supplier's customer.

There is now a growing focus on the end client, being the individuals or organisation benefitting from the use of the building. Even determining this is not simple. With a proliferation of stakeholders – investors, agents, tenants, holding companies, institutional investors etc. – each now has a say in the direction of the project. The move in government to use agencies and project sponsors means a multi-tiered "client" has been created.

To simplify the discussion and place building services into an appropriate context of modern procurement arrangements, two main clients exist. The first is the principal contractor, typified by the main contractor, as most procurement arrangements now dictate that this contractor is responsible for hiring and managing the supply chain. The building service's team is reliant on this contractor to strategically deal with all the issues.

The second is the project client, normally considered to be the party who is the paymaster of the contract. The only time this party becomes the actual client of the contract is either through a contractual relationship with the client, normally under a construction management procurement arrangement, or through a direct contract, either as a sole project or through a service-based agreement.

Their Role

The role of the client is determined by their specific nature. The classification system introduced in Chapter 2 introduced the concept of primary and secondary clients. The emphasis on these ratings should not be understated. Primary clients building for the first time are likely to take a greater interest the building, as a learning process for future developments. A secondary client will merely view the building as a means to an end. This difference will be reflected in the participation level likely to occur.

By assessing the needs an inexperienced client will require considerable guidance through the procurement phase. Regardless of the procurement route chosen the critical criteria that will determine success, and minimise the risk by the client, will be determined by the appointment of a project advisor or surrogate client. By obtaining correct and impartial advice the client's risk is minimised regardless of the eventual procurement arrangement used.

The greatest risk to a client is in the drafting of the project brief. By using modern procurement arrangements such as partnering the brief can be developed as a team with all risks exposed and fully explained to the client. If risk minimisation is paramount to the client then non-traditional arrangements such as design and build or design and manage should be used. Individual categories of clients will determine success by differing criteria and therefore it can be assumed that attitudes to risk must also differ.

Determining the best procurement arrangements for a client is dependent on their specific needs and how these can be best served. The requirement must be linked with the correct procurement arrangement. It is doubtful that an inexperienced client will be able to fully explain or quantify their needs correctly. Therefore, a risk is held by the client of a mismatch between requirements and the executed project. Research has defined typical clients' requirements as:

- function at the right price
- quality at the right price
- speed of construction
- balance between capital and ownership costs
- recognition of risks
- accountability
- innovation of design and technology
- maximising tax benefits
- flexibility to change the design
- the building to reflect the client's activities

- minimising future maintenance
- the need to keep the existing building operational
- a desire to be actively involved

The Principal Advisor

Building services play a major part in determining the viability of both the business case and function of a building. With the demands set by sustainability and energy efficiency the proper installation, operation and maintenance of the building services are critical in achieving green targets. These two issues have considerable impact on the viability of any business to operate successfully, both functionally and economically.

The report *Construction Procurement By Government; An Efficiency Unit Scrutiny* states:

"...we were told, and we saw for ourselves, that problems frequently occur in the later stages of a project, on the mechanical and electrical services and on fitting out. More time and effort spent on getting that right should clearly repay itself many times over. Departments should give early attention on building projects to putting the right team in place to ensure proper planning, design and co-ordination of the M&E services."

Given the above statement, it becomes apparent that the overall strategy for building services provision will determine the success or failure of a project. However, few traditional advisors possess detailed knowledge of the complex engineering and operational requirements that need to be considered during any business case development. As awareness grows of the impact that a buildings operation has on the environment, it demands a greater role for the facilities manager during the project inception.

All of this has created the role of the principal or client advisor. This independent person acts as a unbiased advisor assisting the client in developing a suitable business case and execution strategy for a project.

Construction is not the only consideration. Others include obsolescence within existing buildings and the environmental risks posed by the selected technology. Furthermore, the internal environment plays a major factor in the social well being of employees and their performance.

Principal Advisor's Responsibilities

The principal advisor's role is to set up for the client a strategy and management framework to properly control, monitor and audit the project delivery process. The advice provided must be independent and strategic. If not, then true advice is not being given and the principal advisor is compromised by the criticisms normally levelled against the advice offered by consultants and contractors.

Specifically for building services their main tasks include assisting in the preparation and understanding of the principal project issues that will either affect the economic soundness of the project or the procurement strategy. These will include:

- the business case
- options appraisals, including plant supply and fuels
- industry methods and operations
- initial risk assessments
- requirements under the Construction (Design and Management) Regulations 1994
- the facilities management strategy
- the procurement strategy, including the impact the main project strategy will have on building services
- the contract strategy.

9.2.2 The Designers

The Market

At the time of writing the latest BSRIA market report showed building services consultancy market was valued at £720m. Approximately 2,200 companies, of which the vast majority are either sole traders, or companies employing fewer that 3 staff, execute this workload. However, the top 4% of consultancy's typically have a financial turnover greater than £5 million, and employ nearly 50% of all building services consultancy staff.

Mechanical services account for 56% of a consultancy's workload, with electrical services accounting for 38% and other services 6%. This split in workload highlights the difference between general engineers and specialist consultants. Typically general building services engineers provide advice and design on:

- traditionally mechanical services
- general lighting
- air conditioning
- high / low voltage distribution
- energy
- public health.

A cross-over between general service engineers and specialist engineers occurs with the specialisms of :

- controls
- natural ventilation
- thermal storage.

Specialist engineers normally exclusively handle the subjects of :

- communications
- information technology
- lifts/transport systems

- fire engineering
- acoustics.

Methods of consultant procurement are traditionally equally split between fee bid and negotiation, although with the recent advent of partnering, some 11% of procurement is carried under this practice.

Consulting does prove profitable, with 84% of consultants stating a yearly profit greater than 3%, with 60% claiming profits of greater than 8%.

The Designer's Role and Responsibilities

Central to the principle of determining the engineer's roles and responsibilities for a particular project is understanding the design function within the procurement strategy. The function undertaken by design engineers is required on all projects, but it is not a requirement to have an independent engineer to undertake it. Any number of organisations, including contractors, could undertake the design function, as most are able to provide a composite design and install service.

The decision that must be made within a procurement strategy is whether the design engineer is to act independently of construction or whether an integrated solution, where design and construction are combined, is preferred.

Further considerations must be made as to the complexity of design. Certain installations, such as hospitals, require considerable knowledge and design resources. Few contractors have sufficiently large design resources, both in skill and quantity, to devote to a single project. It must also be considered as to the overall quality of the installation required. Whilst separate design and build may not provide the most cost effective solution, ensuring the design is complete prior to tendering allows it to be fully checked and modelled.

It is common for contractors to subcontract out the design portion. Subcontracting design must be carefully considered. While it allows full integration of design and construction, together with the matching of design skill with construction resources, the control over design can become further removed. This may cause problems in either control or quality if the design team is not fully known at the time of awarding the contract.

The three common assignments of design responsibilities are:

- a consultant engineer
- a contractor with internal design abilities
- a subcontract either through another designer or contractor.

The client can engage the design element through the procurement routes of:

- a direct contract to the client
- a design and construct package
- the main contractor, via a subcontract package.

If an independent design engineer is engaged, then design can take one of three forms:

- a performance specification written by the design engineer in which the contractor develops a full design to meet the stated requirements;
- a partial design, with the contractor engaged to complete the detailed design;
- a full design, with the contractor engaged only to install, with minor design responsibility for elements such as valves.

The project and the procurement strategy determine specific roles and responsibilities. Once the overall strategy for design has been agreed, then the individual responsibilities for each party can be determined.

For the design and completion of a building to be successful it must be a carefully managed process. The designer would normally take responsibility for the design of the visual, thermal, and acoustic environment within the building, through the detailed provision of lighting, heating, ventilation, air conditioning and electrical systems. The role of the designer may not be limited to the design and provision of just the design, depending upon the project procurement strategy it may include the management of the whole project.

Determining the extent of work to be undertaken by the designer will depend upon the overall project procurement strategy. The largest engineering practices can provide an array of services comparable to any multi-discipline design practice. In addition to the traditional design role, three major services now dominate the modern practice: detailed thermal modelling using advanced computer techniques, energy conservation and sustainable design, and commercial services. Commercial services range from initial financial cost studies to detailed whole life cost assessments that are often used in PFI or similar business cases.

The Chartered Institute of Building Services Engineers listed a number of services typically offered by consultants. The more commonly used are listed in Table 9.1

Acoustics	Fire Engineering	Railway Power Supplies
Adjudication	Health and Safety	Research
As Installed Drawings	High Voltage	Refrigeration
BMS and Controls	Infrastructure	Satellite Antennas
BREEAM Assessment	Insurance Investigation	Solar Energy
CCTV Installation	Lifts / Escalators	Sprinkler / Fire Systems
Combined Heat & Power	Low Energy Design	Steam / Compressed Air
Commissioning	Maintenance	Swimming Pools
Computer Modelling	PFI advisor	Television and Radio
Cost Control	Photovoltaics	Windfarms
Energy Audits	Planning Supervisor	
Environmental Impact	Public Health	
Expert Witness	Portable Appliance Test	
Facilities Management	Public Health Services	

Table 9.1 Services Undertaken by Building Services Engineers

Traditionally, the designer's role would commence with an appointment following some form of selection process. The first stage of an appointment would commence with an initial client briefing, where the overall project objectives are discussed in the context of engineering concepts and cost plans influence the viability of a project. For most projects, it would be normal for the designer to work with and provide specialist advice to the other members of the project design team including architects, quantity surveyors and structural engineers.

The output from this process would be the preparation of detailed drawings, specifications and reports to enable competitive tenders to be obtained, or the works negotiated with selected contractors. Although the tendering would normally be orchestrated by the project quantity surveyor, the designer would be responsible for assessing the tenders for both technical compliance and value for money.

Once a project had commenced, the designer would inspect and monitor the installation works to ensure that quality standards are being achieved, through technical compliance. This would also include the review of a detailed working drawings and calculations that are normally undertaken by the contractor. Advice on costs and programme as the work progresses would also be given to the project team. Following completion of the project, the designer could have an extended role in the provision of advice on maintenance, alterations, and actual operation of the system.

Although the above outlines a rather traditional role, the general duties would not change regardless of the procurement role. The biggest variable would be who the designer reports to, and the level or extent to which they are able to communicate with the project team members.

9.2.3 The Contractors

The Market

General mechanical and electrical contractors are the most fragmented members of the building services team. For the UK industry over 145,000 companies share the market, with 78% being sole traders. These sole traders are employed mainly in the housing sector, where they control 80% of the marketplace. M&E turnover was £13.2 billion in 2000, 33% of which was housing, 64% other building and 3% infrastructure.

This work is executed by a workforce that is estimated to be over 260,000 strong, roughly equally split between electrical and mechanical contractors. These figures appear to be surprisingly low as the entire workforce for the UK construction industry is estimated at 1.8 million people, thereby making building services contractors responsible for 15% of the workforce, but claiming 23% of the value. In all, manual workers account for nearly 80% of the workforce.

The largest difference between building services contractors and general construction contractors is evident in the subcontract levels. Where main construction contractors mainly work as managers, building services contractors still rely on traditional directly employed tradespeople. Although larger building services contractors are more likely to subcontract their work, this is mainly due to the use of specialist subcontractors through skill shortages and increased sophistication of technologies involved. On average the larger building services contractors subcontract 25% of their workload.

Interestingly, the manner in which building services contractors are procured appears to fly in the face of common belief. While it is appreciated that small sole traders would normally work for householders or companies direct, in fact the statistics show that 70% of their work is obtained this way. Market research shows that large contractors obtain approximately 66% of their work directly with the client. In fact only 35% of contracts placed with building services contractors are with a main contractor. These figures are obviously an average over the industry as a whole, and vary with the nature of a specific company.

The Contractor's Role

With the changing nature of procurement strategies, the role of the building services contractor has changed considerably in the last 20 years. From being the supplier of specialist labour, modern building services contractors mimic the nature and expertise of most general contractors. Incorporating design, manufacturing and other specialist services, most are capable of undertaking the complete role of the building services team.

It is difficult to draw generalisations of all contractors as each provides unique services based upon expertise, available market, geographical location and corporate strategy. Generally, building services contractors are one of three types, although some sophisticated contractors can operate under all three headings through separate divisions.

1. **Management contractors:** although they are small in number, this type of contractor normally specialises in the overall strategic management of a services installation. Often incorporating the design function, all work, other than the overall management of the project and installation, is subcontracted out to general installation contractors.
2. **General installation contractors:** the most common of all building services contractors, general installation contractors carry out the bulk of the installation, normally using their own directly employed labour. Their work would consist of the normal pipework, ductwork, and plant and cabling operations. Beyond these areas, specialist contractors would be employed on a subcontract basis. This should not belittle their abilities. The diversity of these contractors makes it difficult to give specific listings of their roles. The contractors can consist of single man operations that carry out the vast majority of property maintenance and small installations contracts, to very large contractors employing several hundred people and incorporating a wide variety of expertise. This expertise is split between

their in-house abilities, such as design, prefabrication and facilities management, through to their expertise in different types of installation, such as hospitals, process industries and transportation systems.

3. **Specialist contractors:** although most building services contractors consider themselves to be specialist contractors, most can be classified under the heading of general installation contractors. With the proliferation of technologies and the advent of subcontracting, specialist engineering contractors provide an ever-increasing range of services. Some are specific to particular systems or technologies, such as controls, refrigeration or alarms, where others specialise in particular services, e.g. plant installation, insulation or labour supply.

The building services contractor provides a pivotal role in the management of a construction project. The strategic nature of a project must be broken down into tactical operations that can be executed. The main contractor fulfils this role by working at a strategic management level breaking the project into distinct work packages. The general installation contractor or management contractor must also work on a pivotal basis by understanding what the strategic objectives are for the project, then dividing their own work packages down further to achieve those goals.

The role of the services contractor will be dictated by the procurement strategy developed by the main contractor, or the client under a construction management arrangement. The greatest factor that determines roles is consideration of the various divisions of the building services team. Design and installation can be divided into separate roles, the consequences of which are debated elsewhere, as can the installation into electrical and mechanical services.

Historically the installation contractor also carried out the role of commissioning engineer. Although still largely carried out by the contractor, more sophisticated clients are using specialists to undertake this role in a neutral position between designer and installer.

The Contractor's Responsibilities

The specific responsibilities of a contractor will be dictated by the particular manner in which they have been procured. Traditionally, the responsibility has mainly been split between design and construction. However, this is an over simplification, as design, even under the separate procurement arrangement of traditional construction, is not always clear. Even when the designer supplies a full design, the contractor is still often responsible for minor design elements such as support, builder's work and valve sizing.

As the main contractor procures most building services contractors, specific responsibilities will be dictated by their procurement strategy. Along with the division of design, the installation can be divided into a number of work packages. The final main element of responsibility requiring allocation is commissioning, although commissioning is often an independent operation within the contract. The responsibilities of each party will be dictated by the specific contract used.

The management style adopted by the main contractor will also assist in determining the specific responsibilities to be undertaken by the services contractor. Modern teamwork styles share the specific responsibilities of cost, time and quality on a partnership basis between all parties.

Despite the move by the industry towards more co-operative working systems, the supply chain by its very nature will work on a short-term tactical basis. Whoever is leading the installation, be it one of the three contractor types or a main contractor who has divided the services into separate works packages with different contractors, the main responsibility is to fulfil their contractual responsibilities.

Each subcontract will vary in conditions, even with standard forms of agreements. The three main responsibilities of the subcontractor will be defined under the headings of cost, time and quality.

The subcontractor is normally made responsible for the valuing and cost control of their own works package, reporting its progress on a monthly, or more frequent, basis to the main contractor. Applications for payment based on this account and the valuing of the work executed is the responsibly of the subcontractor.

Although the building services contractor is not responsible for the quality level, as this is dictated by the contract documents, they are responsible for ensuring the workmanship and components used comply with the specifications. This obviously is different if the contractor is responsible for the design element, where they would normally be responsible for converting the quality strategy of the client into specific standards.

Quality is, however, a subjective issue. Much research has been carried out that shows quality standards are determined more by the actions of management and personal attitudes than by contract clauses. It is the responsibility of the building services contractor to instil the necessary empowerment and attitudes within the workforce to ensure quality becomes inherent within the tasks they are executing. This extends beyond the normal site practices and should be determined by the overall attitudes of the company. Quality is normally a constituent part of a company's attitude.

The concept of time and who is responsible for meeting their deadlines, is complex for building services. Under normal construction operations time is the overall period available for the installation of a given set of works. With building services it is often the compatibility between the various work packages programmes that will dictate the suitability of a time period. Given the complexity and inter-relationships between the various systems, careful programming is required.

The programme is quite often beyond the responsibility of the building services contractor and is normally dictated by the main contractor. This does not preclude the services contractor from not properly co-ordinating the work packages within their control. The project programme will set the boundaries within which the services installation must be completed. It is the responsibility of the services contractor to ensure the work is completed within this time.

9.2.4 The Component Manufacturer

Their Role

There is a closer relationship between the building services team and manufacturers than with any other system in construction. The relationship is dynamic, with manufacturers providing a number of key services normally considered beyond the boundaries of manufacturing. This occurs because few building services components are standard, with most major items of plant being purpose made for a particular location. Furthermore, the criticality of a properly planned maintenance regime requires manufacturers to supply a considerable amount of information, beyond the normal installation instructions. Therefore, key manufacturers should become part of the building services team.

Innovation plays a major part in the corporate culture of manufacturers. Most must compete in the global market and therefore are exposed to a greater variety of market conditions and product types. This drives competition on price as well as product features. The end result is that most innovations in both the construction process and in system design are driven by product innovation, as a direct result of the product development efforts of the manufacturers.

Some of the innovation now available from manufacturers has been driven by traditional problems associated with large plant items. Nearly all major items of plant, like air handling units, boilers, cooling towers etc., have some degree of customisation to suit their specific location, or are manufactured to order. This is combined with the declining manufacturing base within the UK, with more components being imported. The resulting problem is extended delivery periods for plant items.

This problem has been partially circumvented by manufacturers adopting modern strategies, including modularisation, standardisation and prefabrication. These strategies are now maturing with the development of both logistics and service-based strategies. More contractors are also adopting similar techniques.

Improved logistics means that previously disparate components are now being consolidated into homogenous products. The result is improved value delivery from the manufacturer, together with a reduction in the number of suppliers required for an assembly. Matched with prefabrication the consequence has been improved project delivery. However, to be successful these types of strategies must be evident within the procurement strategy from the outset.

Manufacturers now play an increasingly important role in providing detailed information on service life, maintenance scheduling and cost-in-use data. As whole life costing becomes more predominant in the business case for a construction project, the need for reliable data on component performance has become critical. Manufacturers are now obliged to provide this information and it is often used as a marketing advantage to demonstrate a product's superior economy or service life.

9.2.5 Specialist Services

Cost Managers

It would be normal in any construction textbook to explain the role of cost manager within the general duties of the quantity surveyor. However, in building services this would severely mislead the reader. Although there are a number of quantity surveyors and quantity surveying practices that specialise in building services it is wrong to assume they all have the necessary specialist skills. Furthermore, it belittles the many other commercial managers, estimators and cost engineers that specialise in this type of work.

The specific person referred to as a cost manager will depend on the stage of the project and context of the role. It is true that the majority of detailed cost expertise held by the industry is with contractors, estimators and commercial managers, whereas expertise in cost planning and whole life costing is normally held by consultants, either engineering or quantity surveying based.

It is normal for the cost management element of a construction contract to be handled by a single person. This is often the cause of problems. There are few people or organisations who possess sufficient skills in both general construction and building services. Other than in the simplest of projects, consideration should be given to the appointment of a specialist building services cost manager.

Cost management is fundamental to determine a procurement strategy. All decisions made within the strategy will have cost implications. Therefore, at the heart of good cost management is the establishment of a realistic budget for the building services work that will allow the various decisions to be made to have their impacts fully assessed. The established budget must continue throughout the project process and its ownership needs to be matched with accountability.

Therefore, the cost manager assigned to a project should posses the following skills:

- be independent, impartial and technically skilled in building services so as to fully understand the impact of various decisions on the established cost plan;
- be able to communicate with the client and demonstrate that project funds have been expended in an accountable fashion;
- have traditional skills in the measurement and estimating of services to provide detailed cost analysis for varying design options.

The specific roles and responsibilities of the cost manger will vary with the procurement strategy adopted for both the main project and the building services. Therefore, it is difficult to provide a definitive list of general duties. However, the following list identifies the main duties of the cost manager responsible for building services:

- to offer advice in the development of the preliminary architectural design development on the impact of the building envelope on services design;
- to provide input into the project cost plan through the establishment of a building services capital budget;
- to offer advice on whole life costs;
- to provide an impact assessment of procurement options on cost budgets;
- to assess the impact of contractual arrangements;
- to provide cost control during the project;
- to provide input into the establishment of the project brief;
- to provide input into or perform value engineering exercises;
- to provide input into the development of specification clauses.

Commissioning Engineers

Traditionally commissioning engineers were the sole responsibility of the contractor. They were often internally employed engineers who would carry out simple air and water balancing for the system. As systems became more complex, however, commissioning became a specialist service. More recently this specialism has developed where it is now separate from the contractor and seen as consisting of specialist engineers in their own standing.

The role of the commissioning engineer can be likened to that of the independent test pilot. Where the designer is responsible for design and the contractor is responsible for installation, all is only theory until someone tries to 'fly' the systems. The role of the commissioning engineer is to provide the independent assessment as to both the quality of design and its installation. This independence provides an invaluable mediation service for the client to identify any problem and to quickly ascertain its ownership and responsibility for correction.

This need for independence dictates that the appointment of a commissioning engineer should be separate from the building services team. Most forms of procurement would allow this to occur. Although separate from the rest of the building services team appointment, the commissioning engineer plays an invaluable part at even the earliest stages of the project and therefore their appointment must be considered as an integral part of the procurement strategy. Independent appointment provides the following advantages:

- the client obtains independent verification as to the appropriateness of design, specification and installations procedures;
- their independence avoids a conflict of interest between designers, contractors and the system performance;
- early identification and resolution of design and installation problems is possible, covering both quality and technology;

- independent verification can be carried out of actual component performance and manufacturer's claimed performance.

The duties of the commissioning engineer are both complex and varied. With specific regard to the procurement stage of construction, their early appointment separate from the building services team allows:

- their participation in establishing an appropriate design philosophy;
- the establishment of an integral commissioning strategy in conjunction with a procurement strategy;
- their assistance in the assessment of designers', and contractors', competency;
- their assistance in plant procurement;
- their assistance in the development of a design responsibility strategy;
- the development of programmes with commissioning logic networks, method statements and quality plans.

The assertion that commissioning engineers should be considered as a separate appointment is supported by the current changes to Part L (Part J in Scotland) of the Building Regulations. Certified commissioning, where the operation of the building must be proven to be within the maximum carbon emission levels is likely to come in to force in 2004. The certifier must be an independent body and therefore for efficiency this role could be combined with that of the commissioning engineer. At the time of writing the full details of how this will be achieved are unknown and therefore the reader is encouraged to seek clarification of this issue.

....

Case Study

Many years ago, we had a job to commission a large bus terminal above a new shopping complex. The bus terminal was reached by a road which dipped through the complex and came out the other side. We eventually got to the point where we had got this set of drawings for the bus station exhaust system, which sounds very simple and straightforward. In fact, when we looked at these drawings it was an amazing design. The consultant should be congratulated because it contained something called a drain extract system, which I had never seen before.

And the way it worked was this. There were some very large axial flow fans with huge aircraft propellers stuck in them. They were mounted up on the roof in a sort of chimney, and that was connected all the way down to the basement of the bus terminal.

Now, we looked at it and we went through the whole thing and we discussed it with the consultant and he said 'Well, actually it's a dual purpose system. It's, first of all, it's an extract system. One of the fans runs and it will ventilate the toilets. In the event that there's an overload of carbon monoxide down in this area, buses being parked, in order to make it completely safe, we've got a carbon monoxide extract system which will actually come into play if this occurs. And if that occurs, a motorised damper will shut off the toilet extract system, the second fan will start and we'll extract then only from the drain areas, carbon monoxide, being heavier than air, will fall into the drains. We'll extract the carbon monoxide'.

So came the evil day that the authorities said 'you've got to prove this, you've go to prove it really works and it does everything it's got to do'. So we designed ho we were going to do that and we tried all sorts of ways, and eventually time ran out and I went to site and I said 'Right. The first thing we've got to do is get a source of carbon monoxide - nice bus exhausts.' Then I asked a really co-operative bus driver if he would oblige me by pulling his bus up so that the exhaust would point straight at the detector mechanisms.

Now, nobody had ever told me that there was an alarm associated with this and it took the form of a huge bell. And after about five minutes of pumping out these fumes, (and I was a bit worried about whether this system was going to work or not), suddenly this bell starts going. And right up in the roof of this bus terminal, I'd never noticed it before, but there was a great big sign illuminated. Nobody told me that was there. Illuminated sign: ALERT, ALERT in red letters. This panics people at the bus stops, who are beginning to think what the hell's going on? We've got a bus going flat out, alarm bells going, a big sign and I've done it. I was concerned, but then I heard the whack of this motorised damper closing. I thought 'Thank God that's worked'. But, of course, straightaway the second stage of the fan started up. Now that actually sounded like a Boeing 707 just coming in above your head. Then I noticed out of the corner of my eye, that to my horror there were pigeons wandering down the road and one by one they were disappearing down the drains. I couldn't believe my eyes.

When I went outside, the mess up the street was unbelievable because, having gone through these fans, which I told you are like aircraft propellers, down the drain and through the fans, it was... The system passed. The system passed because nobody wanted to do it again.

Actual story told by Chris Burgess of Burgess Commissioning at the Rethinking Procurement Conference October 2001.

....

Specialist Engineers and Contractors

The increasing complexity of technology and subsequent specialisation by the industry has been previously stated. The result is a greater diversity of specialist engineers and contractors who provide specialist expertise that is normally focused on a specialist technique or technology. In short they are the specialist specialist.

These specialists vary in the type of service provided and can range from specific techniques, such as cable pulling to engineering, such as solar energy.

The specific specialists involved will depend on the requirements of the project, typically this could include:

- **specialist engineers:** for solar/alternative fuels, energy assessment, utility co-ordination, telecoms, commissioning, acoustics, fire engineering, ventilation, process engineers, steam.
- **specialist engineering contractors:** fire protection, control systems, acoustic insulation, fire/security, insulation, commissioning, operations and maintenance manual consolidators, lift, escalator.
- **specialist contracting services:** cable pulling, plant movement and installation, prefabrication, utility, setting out.

Each of these must be considered within the context of procurement, as to whether it is favourable to allow them to be procured as a subcontracted service by a member of the building services team, or whether it is favourable to procure them as a distinct works package.

The party best able to understand the technology involved will determine the main criteria in which the specialist should be procured. As a simplistic rule, a specialist will supply a service specific to the design, such as controls, and therefore be integrated within the building services team, or may be specific to a clients needs, such as process engineering within an industrial process, thereby the client possessing the knowledge of the technology. The possessor of the knowledge should be the one that sets the procurement strategy and decides the relationship with the project.

As the technology involved will be specific to the contracting company or designer, the risk and liability of performance should be placed on the specialist provider to guarantee their performance and the product.

To be successful the specialist element needs to be managed as part of the total design process, with the specialist's knowledge and advice being made accessible in the early stages of design. Value engineering and buildability analysis must become an integral part of the specialist's skill base.

For specialists involved with delivering particular products to the marketplace, two general business strategies are normally adopted that will affect the manner in which procurement can take place.

Certain specialists will adopt a manufacturing-based strategy wherein to protect market share and obtain an economic volume of throughput, their service range will be limited to the products produced. This approach means that designers will have to work with a limited set of options, therefore selection will be based upon the product rather than the supplier.

The alternate approach is based upon bespoke manufacturing. A differentiated product strategy, based on customisation, requires an assessment and development of a procurement strategy that matches both product and supplier characteristics. This requires specialists to develop an ability to input their skills during design to optimise the design and production process for one-off solutions.

REFERENCES

1. ACE (1995) *Association for Consulting Engineers Conditions of Engagement*, London, Association for Consulting Engineers.
2. BSRIA (1998) *Customer Satisfaction Survey*, Bracknell, BSRIA Publications.
3. BSRIA (1999a) *M&E Contracting - Market Size, Structure, Sectors and Self-employment*, Bracknell, BSRIA Publications.
4. BSRIA (1999b) *Building Services Consultants - Market Size, Structure and Sectors*, Bracknell, BSRIA Publications.
5. Levene, P. (1995) *Construction Procurement By Government; An Efficiency Unit Scrutiny*, London, HM Treasury.
6. Lynton PLC (1995) *The UK Construction Challenge*, London, Lynton PLC Company Report.
7. Masterman, J.W.E, (1992) *An Introduction to Building Procurement Systems*, London, E&FN Spon.
8. Turner, A. (1990) *Building Procurement*, Basingstoke, The Macmillan Press Ltd.
9. Wild, J. (1997) *Site Management of Building Services Contractors*, London, E&FN Spon.

Chapter Ten

Forming the Team

10.1 THE PROJECT TEAM

The importance of the human element within any business management system is now fully recognised. Even within procurement strategies, over the last 10 years, emphasis has shifted from the analytical development of a system marshalled by procedures and contract clauses, to a more philosophical attitude where importance is being placed on selecting compatible organisations that can work as a true team.

The leading idea is based on partnering. But very much like contracts, partnering is a state of mind rather than a precisely defined methodology for procurement. Partnering is based on the notion that all parties that form the project team do so with the mutual objective of completing the project successfully. To accomplish this the project partners must be selected for their soft abilities of trust, co-operation and teamwork. The management systems used to enforce the philosophy must not replicate the traditional divisions, and must be co-ordinated to provide an environment of continuous improvement.

Partnering must not be seen as a cure-all to traditional problems or as an easy route to achieve project success. To work fully it requires even greater commitment from all the parties and a clear strategy as to why such a method is being undertaken and what outcomes are desired. It should never be used in lieu of a properly considered procurement strategy.

To fully understand how partnering can improve procurement, the general nature of how project teams are formed must be considered. From this, the single most important enabler of partnering is good leadership. Leadership determines the strength of a team.

For building services, the team is an eclectic mixture of designers, engineers, technicians and trades-based professionals. Of all the teams operating in construction it is probably the most diverse. Therefore issues such as co-ordination and integration are paramount to achieving a successful team environment.

10.1.1 Co-ordination of the Project Team

The complexity of buildings, and in particular building services, has to reflect the demands of a modern society that is driven by information technology and self-fulfilment. Legislation and the specialisation of technologies has resulted in building solutions that are more complex and require higher levels of specialist contractors.

This increase in technology, together with the increase in more sophisticated supply strategies, has resulted in project teams that require a high level of co-ordination to produce a consolidated product. But increases in both team members and specialists have created their own problems. Each increase in a specialisation of technology creates another paradigm that incorporates its own language, customs and attitudes. The result is fragmentation of the team. With fragmentation comes a breakdown in communication and differences in objectives. All this places increased emphasis on the main contractor to properly co-ordinate the works.

The main contractor acts as the team leader, and therefore their actions will dictate the success or otherwise of the project. Communication is the key. The leader of any team is responsible for obtaining an instruction, then translating it or breaking it down into specialist actions for individual members. Good management skills are then needed to ensure that the assigned tasks are both executed and co-ordinated to form the intended project.

Importance of Co-ordination

The emergence of building services as an increasingly specialised discipline raises the importance of co-ordination on two levels, namely projects and services. The main contractor is usually responsible, (although this can be debated due to specific procurement arrangements and structuring of contracts), for the co-ordination of the various disciplines and major subcontracts works, known as project co-ordination. However, a more detailed co-ordination exercise is needed between all companies, both designers and contractors, to achieve a fully integrated building services system. This is known as services co-ordination and is dealt with later in the text.

Co-ordination not only applies to the physical entities of a project. Each technology is now embraced in its own language, and given the structure and nature of building services, this is dominated by "engineeringeese". This can often cause problems during design discussions with disciplines of a more aesthetic nature. Poor communication results in a lack of a unified decision over a problem, often causing compromises to be made.

Construction Process

Procurement sets out to form a management strategy and system that will enable the three critical processes of design, construction and co-ordination to be successfully executed. Much has been written about the differences in objectives between professionally motivated designers and consultants, and the commercial motivation of contractors. Therefore, the organisational structure of any project will depend upon the procurement arrangement at any point within the project.

The allegiances of individual project members will be divided between the short-term objectives of the project, and the longer-term view of their associated company.

To resolve this conflict, a move has been made towards procurement arrangements that are based on keeping project teams together. Framework agreements and partnering arrangements dominate these strategies.

Nature of Specialists

One of the major issues that appear to affect the manner in which success is achieved in procurement, is the specific nature of business operations undertaken by specialist contractors. Specialist contractors are divided into two types:

- **product orientated:** these respond to a performance specification with a product in which a design is incorporated. This can mean that installations are design orientated rather than product orientated. Fragmentation of work into distinct packages is leading to increased co-ordination problems, particularly on major projects.
- **systems orientated:** these either carry out a design from inception, in response to a performance brief, or develop another basic design into a working detail.

These business classifications must be contrasted to the management style of the company. The construction industry is dominated by two forms of management: either organic – responding to their environment, like main contractors; or mechanistic – in the form of specialist contractors that work in a stable supply chain, with known environments. This could possibly lead to an incompatibility in procurement strategy as the project may be managed by an organic system trying to adapt and utilise mechanistic supply chains.

A main contractor may require an organic approach with high integration and low control to create flexibility. Research has shown that for co-ordination by a main contractor to be successful it needs to be organic as a mechanistic paradigm usually has a poor success rate.

Within procurement it is usually believed that the abilities of the main contractors to procure the correct specialist contractors is due to their strategic abilities of dealing with the supply chain. It is normally assumed that the main contractor holds a dynamic relationship with the market, with the consultants determining strategic objectives from the client.

In reality, however, main contractors are management contractors sub-letting almost all the construction works to domestic subcontractors on documentation that does not recognise this to be the case. However, this assumes the subcontractor has no strategic influence on the project and will make the best possible choice in terms of resources and methods. Furthermore, the critical factors for project success are compatible with the delivery process adopted by the subcontractor, i.e. what determines project success may not equally determine a successful building services installation.

The main differences between management and traditional contracting include the contractual distribution of risk and to a lesser extent, the method of price determination. The result is a main contractor with little strategic control over risk or price on the projects, as their role has become that of a co-ordinator only.

Co-ordination and Integration

Due to its dynamic nature and interoperability, building services must be fully integrated and co-ordinated for success. The success of co-ordination, and therefore that of the project, is related to appropriateness of a particular procurement system.

Co-ordination is the act of achieving integration by placing items in their correct position relative to one another and the systems to which they form a part, with integration being defined as the state of incorporation.

In the construction industry procurement arrangements are very often open systems, in that they respond to their environment and are essentially a transformation process. In this way the primary actors in a construction project, namely consultants, the client and main contractors operate in an environment that is highly permeable, requiring constant change. The main project strategy is divided into five subsystems: goals and values, technical, structural, psychological and managerial.

The main purpose of the five subsystems is to transform the resource inputs into the desired outputs, while at all times responding to their open environment. Overlaying this idea of subsystems is that of organisational boundary models, and it becomes clear that there is a link between the organisational subsystems and the strategic paradigm. Therefore, the structural and managerial elements must embody the differential and integrative aspects associated with co-ordination. Contingency theory believes that to be successful, equilibrium must be held with the environment. To ensure success, this ideal must hold true for the entire supply chain involved with a project.

Interdependence though is understood and has been previously well researched. Reciprocal interdependence is where the output of one system becomes the input of another, very much like the supply and technical aspects of building services. These are the most difficult to integrate and require considerable co-ordination and co-operation, thereby highlighting the importance of co-ordination to building services. Similarly subsystems are differentiated, whilst simultaneously requiring high levels of co-ordination and integration to achieve optimum performance.

All of this confirms the belief that co-ordination of building services during construction is the most demanding of management tests. The influence of a procurement system on successful integration and co-ordination should not be dismissed. To achieve true integration, it is necessary to use a procurement system conducive to the effective co-ordination of building services. However, the successful outcome of a project is not necessarily linked to procurement arrangement per se, but rather the final relationship between members of the project team.

Co-ordination must respect the client's brief, conditions of engagement of consultants and contractors, division of design responsibilities, allocation of risks, early incorporation of specialists, quality of design, construction management and form of contract. Although previous research has shown that no general consensus of factors is needed for good co-ordination, specific factors that affect co-ordination, include:

- engagement conditions and fees for consultants
- communication
- design procedures and practice
- management structure
- cost effectiveness
- site organisation.

The requirements for integration include:

- fabric designed while minimising thermal requirements
- integration of services and structure
- a design suitable for practical realisation
- proper monitoring of design and construction
- a motivated team
- clients demands being met.

It appears that part of the problem with co-ordination stems from contractual frameworks that are often incompatible with the needs of the procurement strategy, and critical success factors needed for project completion. Information co-ordination is specifically excluded from the Association of Consulting Engineers (ACE) standard form of agreement for fees structure, unless an additional payment is made to the consultant.

The RIBA *Conditions of Engagement* state: " *The Architect will advise on the need for independent consultants and will be responsible for the direction and integration of their work*". Yet in the ACE Memorandum of Agreement, by which most services consultants are hired, no statement of responsibility for integration and co-ordination exists. Nor is there provision for co-ordinated drawings.

10.1.2 Formulating the Project

Temporary Management Organisations

The project procurement strategy concentrates on the formation of a team, as a singular industry to complete a project within a given environment. All projects are based on a temporary management organisation (TMO), as they are abandoned at the completion of every project. But while this holds true for the primary actors, the specialist supply chains normally stay together and obtain other projects within their general portfolio of work.

Building services is largely an assembly industry, where the skill and knowledge of each technical system is held by a number of unique specialist companies, consultants and component producers. Supply chains are well established and although temporary, operate within a determined environment, and therefore display some of the characteristics of a mechanistic system. Furthermore, each technical system has its own unique suppliers, contractors and methods of operation. Thus the building services industry can be seen as a complete system, but highly influenced by its own environment and stable subsystems.

The TMO, then, must formulate a strategy to engage the supplying industries, represented by specialist engineering organisations, to act in a dynamic manner to deliver the project results. The integration of and co-ordination between, supplying industries will determine the ability of the TMO to act as a singular industry. The process used to link the environments, and the compatibility between the success factors of the supplying industry and the project, will determine the success of this interaction. Therefore project success will be determined by the interrelationship between critical success factors and processes.

A building services installation is undertaken through a diverse supply chain, based upon design responsibility and technology. In essence they supply a holistic product, through a mechanistic supply chain, that provides a distinct functioning system. This is in contrast to the organic relationship a main contractor has with the project environment.

Traditionally, building services contractors were part of the TMO, through the nomination process, and were contractually bound not only to the main contractor, but to the project, with the client and consultants each having specific duties. With the decline of the nomination process, however, two forms of procurement now largely exist for building services: direct procurement as a domestic subcontractor, and therefore not part of the TMO; or being the primary partner through either a Joint Venture Agreement or a Prime Contract.

Two hypotheses dominate the field of procurement: firstly, success is determined by the manner in which procurement form is managed, i.e. the management structure and actions are more important than the procurement arrangement; secondly, the nature and structure of the TMO determines the appropriate procurement form. Both of these ignore the idea that the supply chain is involved and assumes the TMO has direct influence over its environment and the co-ordination/integration of the players.

Clients are adopting more sophisticated supply strategies to their non-core sourcing activities, such as component supply, energy, and facilities management and servicing, causing an increase in service-based supply strategies for building services. Therefore the business case development for projects is now extremely complex, requiring considerable input from the client. This developed project strategy must then become inherent within a developed supply strategy.

The manner in which the project TMO functions has direct influence on the outcome of the project and the installation of building services in particular. High levels of value require special attention and organisation to be successful, therefore intimating that certain project elements may be detrimental, and may not be limited to just buildings services.

Social Structures

The importance of social relationships affecting project success has been well documented in the context of procurement. However, very little exists on the subject of building services.

The organisational structure, a subsystem of the procurement system, formalises the relationship between technical and social subsystems. The social subsystem consists of shared beliefs, values and a network of social roles and relationships, and is often imported from the external environment. Within building services, this could mean that each services system or industry sector could have its own unique social structure and therefore paradigms.

The external environments of the individual subindustries or systems often influence these same social structures. These are personal influences from such things as an individual's role in social units, clubs, families or associations, with these being carried into other contextual environments. Not only are these social systems specific to individual organisations, but they reflect the development over time of values and attitudes.

Furthermore, technology affects the inputs to a system, whilst the social system determines efficiency in which they are used; and technical subsystems are also determined by the nature of the work carried out by organisation.

Therefore, the various differences in subsystems must be resolved in the formal structure of a construction management system. Managerial subsystems span entire organisations and must relate it to the surrounding environment, and can be generically divided into three levels of management:

- **technical:** strives for rationality and the economic use of resources
- **organisational:** balances short-term technical needs with long-term institutional ones
- **institutional:** relates activities to the environment.

Project Organisation

To ensure project success, the client, either independently or with the assistance of a principal advisor, must organise a large number of individual organisations into a team using a process of organisational design. Utilising this with systems thinking to define a client's objectives, as being determined by the environment in which the client operates, can be used to define the objectives of using a particular procurement route. The definition of an organisational system assigns specific responsibilities and authorities to people and organisations, and defines the relationships of the various elements in the construction project.

The very idea that a construction project can be divided into appropriate works packages can be criticised, as construction teams are multi-disciplinary, cutting across organisational boundaries and are often one-off arrangements set up to tackle a specific non-routine but difficult problem. It is believed that this does not hold true for specialist contractors. It may also be that a hiatus exists between the temporary team of the project, who are not used to design and build, and the rather semi-permanent teams held within the specialist engineering industries who regularly work as a design and build organisation.

The organisational culture could be used as a basis to define objective requirements of the environment, as many ills within organisations stem from imposing an inappropriate structure on a particular culture, or from expecting a particular culture to thrive in an inappropriate climate. Furthermore, organisations are either power-role-task or person-orientated, and suitability of a particular form is determined by an organisation's size, ownership, technology goals and objectives, environments and people; therefore a procurement system consists of matching people to systems, tasks and environments and of interrelations between all four.

Critical Success Factors

The importance of specialist engineering contractors to the construction process has been widely stated. The role of the main contractor now exists only to co-ordinate the work and act as a general operations administrator. The work is executed, as is the supply of all resources, together with an increasing amount of detailed design work, by specialist contractors. Therefore, it would be reasonable to expect that decision models should place some importance on this notion.

However, successful procurement demands a detailed understanding of the role of the specialist. For the building services to be successfully and adequately incorporated into building a project, the choice of the correct procurement system for a particular project is of paramount importance. The project environment will highly influence, and be influenced by, the nature of the specialist engineering industry. The subcontractor is now the controller of both resources and knowledge and therefore will directly influence the technology (by possessing the knowledge and skill required) the project (by determining the response to the complexity) and the environment (by largely determining the environment of the construction industry).

In most writings, the critical success factors for any construction project are often given as: communication, project mission, top management support, project schedule or path, client consultation, personnel recruitment, technical task, client acceptance, monitoring and feedback, and trouble shooting. The current trend to more co-operative working and focus on management soft issues has added the critical success factors of an understanding of team interdependence, cohesion, trust and enthusiasm. Finally, recent research has highlighted the importance of conflict and resolution, management of TMOs, team flexibility, team control and administration.

These identified critical success factors must be compared within each party since each has its own project objectives and criteria for measuring success, but common to all being on time, working within the budget, maintaining profit goals and avoiding legal claims. These can then be translated into critical success factors for project success: a cohesive team; contracts encouraging team behaviour and allocating risk fairly; experience in all project phases; designability; constructability; operability; and information being available in a timely manner. Two key factors begin to emerge, the first being no legal claims, the contingency view of procurement systems beginning to emerge. The second is the concept of success, and processes critical for the accomplishment of the brief, including the supply chain.

Procurement is a system in which five subsystems operate which, when arranged in their contractual sequences, give the generic procurement arrangement options:

- works packages: either horizontal or vertically split
- functional grouping: separated, integrated or management
- payment modalities
- contract conditions
- selection methodologies.

The selection of available options for each subsystem, and the development of a suitable procurement system, is valuable to clients in increasing the likelihood of project success. In the same light, appropriate works packaging, which can be identified as a subsystem, is critical for success. Although this may favour a procurement route based on works packaging, namely construction management, this method is often criticised by specialist engineering contractors for construction managers breaking the project into indiscriminate lumps, requiring more co-ordination or disjointed working.

The idea of works packaging must be contrasted with the nature of the construction industry and respect its business structures, for construction is a multi-industry in nature, which possibly contradicts the required nature of a TMO to be successful. The procurement problem, therefore, consists of finding the best "design" for the TMO, given the particular building client and their context, such that the various task-organisations who will be called upon to work on the project give their best performance.

10.1.3 Team Hierarchy

Leadership

People hold the key to improving our industry. There is currently a wide discussion as to how changing processes and product delivery mechanisms will meet the demands of a modern industry. Improved performance relies on a motivated and valued workforce, working in a co-operative culture of continuous improvement.

Sir John Egan's report *Rethinking Construction* set the agenda for the industry's problem with its stated reforms. The five key drivers for change were described as:

- committed leadership
- focus on the customer
- product team integration
- quality driven agenda
- commitment to people.

These lead on to four key areas of improving the project process:

- product development
- partnering the supply chain
- project implementation
- production of components.

Essentially these address the four critical Ps of any business - people, process, products and purpose.

The building services industry for a long time has invested in product development and the production of components. It must now focus and draw its attention to the two main issues of partnering the supply chain, and project implementation. To achieve these building services must address its traditional place within the supply chain, and the skills possessed by its managers.

In Europe it was recognised that the technique of quality self-appraisal is very useful for any organisation wishing to develop and monitor its performance. The European Foundation For Quality Management has developed a business excellence model that can be used by a company for a systematic review and future strategy of its people. The EC business excellence model recognises that processes are a means by which a company or organisation harnesses and releases the talents of its people to produce results.

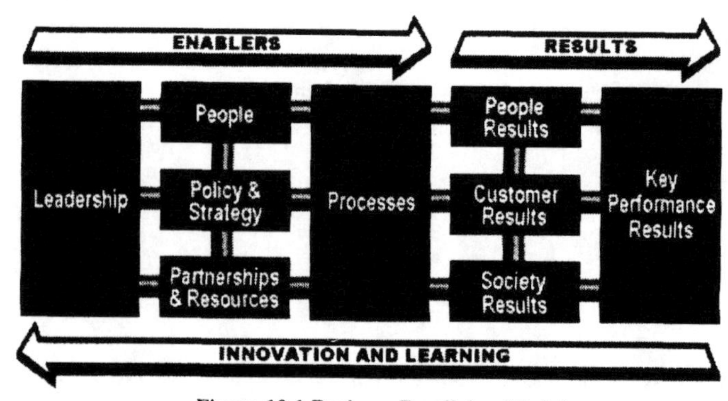

Figure 10.1 Business Excellence Model

Furthermore, the processes and the people are the enablers that produce results. The business excellence model is read from left to right, showing that leadership from senior management is needed to enable improvements in people management, policy and strategy, and resources. Senior management must provide the necessary vision and values that promote improvement. Clear goals and purposes must be given through policy statements.

All of this leads to improvements in the process. These result in people satisfaction and customer satisfaction, both of which have a positive impact on society. The final output is improved business results.

Although the linking of people to a quality model may at first appear abstract, the key concept in improvements is quality. This does not merely apply to the product or services delivered, but to the entire management chain. What dictates competitiveness in the general marketplace is the efficiency in which operations are undertaken. In a typical building services installation the amount of hard resources and labour required will be static between all competing companies. What will dictate the efficiency, and thus profits, is the competence with which the resources are managed and the innovation deployed.

An organisation's ability to function and prosper depends, in a large part, on the knowledge and skills of its people, and the knowledge base that it collectively holds and deploys. This knowledge base includes any information, however hard or soft, that contributes to the organisation's operations and success. The agility and impact with which the knowledge base can be leveraged depends on the quality of the knowledge system in place.

Knowledge has become a commanding business issue for a number of reasons.

- The shift away from capital assets as the basis for market supremacy in favour of knowledge-based tangible assets. The traditional reliance on engineering and technology as the basis for a good business no longer holds true.
- The emergence of a number of technologies capable of capturing, managing and disseminating vast quantities of information. Already the internet plays a massive part in most businesses.
- The move towards the virtual organisation, where boundaries become blurred through the use of alliances, strategic partnerships and outsourcing relationships. In these, the relationships between an organisation and its people change to accommodate new ways of working making it imperative for organisations to manage and capture vital knowledge. This is a critical issue within the building services industry and the true future for building services lies in delivering joined-up systems and networks as fully functioning building systems.

....

Case Study

The United States Marine Corp

One organisation that has similar problems to that of construction is the military. They are required, like construction, to do anything, anywhere, at anytime. They must have the ability to quickly survey a situation, understand the objectives and complete the mission. Often this will require different leaders throughout a mission as various phases require different skills.

The Marine's discipline is based on five practices, backed up by three key values of mission, value and pride; all of which are applicable to construction.

Practice 1: Overinvest at the Outset in Instilling Core Values

Recruits spend their first 12 weeks understanding the core values of the Marine Corp. The values of honour, courage and commitment are instilled by a cleansing process where each recruit must state all aspects of their past. This cleansing is done publicly, to ensure all people within the unit are equal. This requires true teamwork as any exercise not completed by a single member, must be redone by the entire unit. Therefore, the stronger members assist the weak, raising the general output of the unit.

Practice 2: Prepare Every Person to be a Leader and a Follower

Unlike businesses where only the brightest are taught leadership, every Marine is given training in both leading and following. This creates immense pride, mutual respect and loyalty as a unit. Each Marine can rely on his comrades for support and backup. Leaders are not produced as a stereotype. It is believed that a variety is required, from assertive to collaborative, insightful and supportive.

Practice 3: Distinguish Between Teams and Single Leader Workgroups

A team is characterised by a shifting leadership, undertaken by the best person at any particular time. They collectively decide the goals, based on set objectives, and accept responsibility for failure as a whole. Single leader work groups have a leader who designates responsibility to others. Goals and performance are assessed by the leader. Responsibility for failure is taken by the person whose actions failed the group.

Principle 4: Attend to the Bottom Half

The Marines sum up their attitude in *"You may give up on yourself, but we will never fail you."*. Despite their movie image stereotype Drill Sergeants will go to extraordinary lengths to ensures the weakest member of the unit is brought up to standard. Before recruits show up at training camp, Drill Sergeants will study photographs and files to ensure they know everyone by sight.

Principle 5: Use Discipline to Build Pride

They use the core values of honour, commitment and trust, to ensure each person is self-disciplined, but also ensure that people around them act with equal commitment. The recruits quickly discover what they can accomplish if self-control is exercised. They then learn that collective discipline within a group can accomplish even more. As they are judged as a group, a collective drive is formed to complete a mission.

Although the business excellence model provides a starting point for corporations, personal improvement is required. Robert Kelley in his book *Star Performer*, which he based on research into the top performing employees at Bell Laboratories, concluded that to become personally successful, people should posses the skills of:

- **initiative**: being willing to work outside their given tasks to solve corporate problems;
- **networking:** using personnal contacts to solve problems;
- **self management:** managing all aspects of both personnel and business requirements;
- **perspective**: understanding the main company goals and ambitions;
- **followership**: described as "checking your ego at the door", willing not just to lead the team, but to be lead by others;
- **leadership**: leading the small projects as well as the headline ones;
- **teamwork**: actually doing it, not just paying lip service;
- **organisational savvy**: using street smarts to tackle political problems;
- **show and tell**: being able to persuade audiences and deliver the correct message.

....

Case Study

The 4-Hour House

What can leadership and changing the culture of the coal face workforce actually achieve?

The best example is given by the San Diego Housebuilders Association. To prove their ability to deliver quality houses in realistic times the idea of the quickest house was born.

Three months of planning, designing for erection and numerous gatherings of workforce to discuss roles, programmes and activities culminated in a world record attempt. The first year was trial a session. Building a three bedroom, double car garage bungalow in timber frame normally takes about three months in America. This includes all foundations and landscaping. The first attempt saw the house completed in 4 hours 18 minutes.

The subsequent year, together with minute by minute programming, resulted in a house being built in 2 hours 45 minutes.

A written description does not do justice to the leadership required to direct 350 men and women to accomplish such an event. However, four key aspects must be respected to attempt such a feat:

- designing for construction
- planning, planning and more planning
- leadership directed to true teamworking

- motivation, motivation and more motivation.

People make an industry. Quality people, motivated to work efficiency will enable all organisations to prosper to form dynamic industry. Providing strategic leadership while allowing people to exploit their full knowledge raises satisfaction not only for employees, but the customers they contact.

....

10.1.4 Partnering

The recent government reports have highlighted that clients should embrace a more team-spirited approach to construction, thereby raising the profile of partnering with construction. The concept was noted in the publication by client-based organisations who used it within their own industries. *Rethinking Construction* has further challenged existing construction practices by stating that *"the industry must replace competitive tendering with long term relationships..."*. Both reports share the ideal that a more integrated project process, where parties work together for the good of the end product, is the only way of achieving a successful construction project.

Partnering is not a precisely defined concept. A number of similar ideas overlap creating an abstraction of similar concepts. This has caused an array of expressions used to describe the general idea of partnering. These vary from formalised principles that are reflected in procurement strategies, such as alliances and framework agreements, to more general philosophical views on the manner in which individual parties should behave. Therefore, no precise definitions exist, rather a general set of ideals exist to which partnering agreements should aspire.

The definitions that do exist can be generally characterised as follows:
"Partnering is a management approach used by two or more organisations to achieve specific business objectives by maximising the effectiveness of each participant's resources. The approach is based on mutual objectives, an agreed method of problem resolution and an active search for continuous measurable improvements".

The objectives set out within this definition can only be achieved through the building of long-term relationships and collaborative team-working, both for the success of the project and for the benefit of each project partner. These must then be underpinned by the following critical success factors:

- commitment towards co-operation and team building
- trust
- identification of mutual objectives for both the project and each other's business
- dispute resolution procedure
- openness in communication and financial transactions
- senior management involvement and commitment
- benefit sharing

- accountability and fair allocation of risk
- continuity of work and personnel
- early involvement of all key players
- honesty
- shared resources.

The consensus of opinion shows that even greater benefits can be made from strategic partnering. This is where clients and their suppliers work together on several projects, in contrast to the more common project partnering where the relationship between individual partners only lasts for an individual project. Strategic partnering provides a better framework for continuous improvement, including cost reductions, improved productivity and profitability.

Current thinking identifies the four key critical success factors of partnering to be:

- commitment towards team working/a synergistic approach to the project
- trust
- identification of mutual objectives
- communication-agreed plan for timely (lowest level) dispute resolution.

To be successfully integrated within the project environment these general principles must be inherent within the procurement strategy. Although the use of modern contracts and a more open approach to communications will assist in developing the necessary co-operative environment, partnering is a philosophical subject, and therefore will only be successful if each member of the partnership understands the underlying ideals.

Commitment

Establishing commitment is a form of communicating each firm's roles and intentions unequivocally. Communicating commitment reduces client and competitor uncertainties and enables them to make strategic plans from new assumptions. This point is a key driver for embracing partnering within the context of procurement. Greater commitment by both parties to long-term relationships and to each other is more likely to heighten trust between them.

Commitment can be demonstrated in many ways, and transcends the assurance of a forward workload. Making a conscious decision not to take any damaging action at the time of a disagreement is a form of commitment that creates trust. A persuasive way to communicate trustworthiness is for each firm to demonstrably take responsible actions (e.g. sharing knowledge, adopting a non-adversarial approach) that benefit the whole team.

The importance of commitment should not be underestimated, both managerially and economically. To make partnering work requires a high commitment of all parties' time. This is necessary to ensure that communication lines remain open and members are focused on the objectives. On a purely financial basis, consideration must be given during the development of the strategy as to the additional labour input required and how this is balanced against other benefits derived from partnering, such as reduction in overall costs or minimising the cost of litigation.

Trust

At its ultimate level, trust is the unquestionable belief that each partner can rely on all other team members. In practice, given the past adversarial attitudes, the difficulty in making this leap of faith is acknowledged.

Moving the parties of a team that have traditionally worked in the manner of self-preservation to one of trust and teamwork is not easy. Better understanding of each partner's objectives and roles, commitment to the team and eliminating the likelihood of adversarial practices, all work towards having a trustworthy state. This must begin at the procurement stage, by careful selection of compatible team members. Furthermore, efforts also need to be made to ensure the entire supply chain is included and selected on the balance of quality, value and commitment.

Identification of Mutual Objectives

At the outset, each partner should have the opportunity to state what they expect from the partnership and what contribution they can make to the team. Great importance should be placed on having a better understanding of what each partner's objectives are, both for the project and as a business. These must then be considered with the view of establishing a mutual aim for the project and a core set of objectives for the team can be established. This will often be documented in the form of a mission statement.

The important issue in setting objectives is to ensure that they are mutual to the entire team. The benefits from having a more open, cohesive, relationship must be realised by each member of the team and not just benefit the client. Unless the objectives are mutual then buy-in by individual members will prove difficult.

The Client

The end-user client is at the core of any procurement strategy. To truly achieve the benefits of partnering the client must be of a similar mind and be willing to commit to the additional resources required.

It would be logical to assume that the client must be a frequent buyer of services and be experienced in selecting bidders, in order to appreciate the benefits of using the synergistic type of management that is availed through partnering. Yet 80% of UK construction is undertaken by A1 secondary inexperienced clients, who will rely heavily on a professional team. Therefore, most clients will require good, impartial advice to guide them in making such a decision.

Partnering has evolved in the private sector but there is no organisational reason why a public sector purchaser should not employ the partnering concept as a technique to obtain better value for money from its suppliers. In fact in the UK, the government's current Best Value initiative is aimed solely at achieving increased value for money in procurement through the use of partnering. There may well be legal obstacles, in so far as the practice is opposed to the requirements of the EC Directive on Procurement, but at the time of writing moves were being taken in Europe to eliminate or curtail these barriers. If the sole criteria for award of the contract is lowest price, then the concept of partnering, by definition, is excluded. However, once a contract has been awarded, there is no provision in the current Directives, that would prevent the parties involved working together to improve the delivery, quality, etc. Negotiated procedures can also be used where a new project consists of the repetition of similar work carried out within the previous three years, in which case, no advertising for open competition is required. There appears to be sufficient flexibility in the Directive for partnering to be a consideration for certain projects.

Dispute Resolution Procedure

Given the past adversarial practices throughout the construction industry, a major factor critical for successful partnering is having a systematic approach to problem resolution. The aim is to seek "win-win" solutions, rather than a party to blame. The process for dispute resolution is based on attempting to resolve problems as they arise, aiming to find a solution promptly at the lowest level possible. There is evidence that demonstrates that partnering arrangements have fewer disputes than traditional procurement methods.

It would be foolish to say that disputes never exist in partnering arrangements. Partnering benefits parties in dispute by making sure each understands their respective objectives and that of the project. A dispute escalation procedure can often ensure that minor disputes are expeditiously resolved. Some believe that the principles of partnering can be applied to a troubled (initially non-partnering) project because partnering is a corrective action process.

One of the key mechanisms in partnering arrangements is having a clear procedure for dispute resolution. The philosophy is that reducing the likelihood of conflict benefits every partner – hence the "win-win" buzz phrase.

Research conducted by BSRIA showed that in 50% of cases where the client/main contractor were partnering, the building services contractor was not included in the partnering arrangements. Given that the building services element comprises, typically, 33% of a total new building cost and nearer 47% for a refurbishment project, neither the client nor the specialist contractor are ideally placed for a win-win outcome.

However, if partnering is used with traditional procurement arrangements then a perpetual conflict will exist between design and installation. A procurement arrangement is needed that allows early appointment of both consultants and building services contractors to define roles, responsibility, risks and the sharing of knowledge and would enable each party to work towards a win-win situation.

Along with the appropriate procurement arrangement, a number of mechanisms have been identified that assist the partners in achieving a win-win situation. Most are fairly innocuous, in that they are simple and easy to undertake, providing benefits beyond partnering. Workshops and value engineering exercises allow the building services and main project team to meet, thereby breaking down communication barriers and allowing individuals to appreciate the project from another's perspective. They provide:

- clear procedures for conflict resolution
- a formal partnering charter (mission statement)
- a means of measuring project objectives
- setting performance benchmarks
- independently facilitated workshops
- project evaluation and debriefing
- value engineering
- commitment to continuous improvement , such as using TQM
- off-site social get-togethers
- joint skills training.

Contractual Issues

Partnering raises an interesting philosophical question – is there, or is there not the need for a contract? All partnering agreements should consist of a partnering charter where the objectives for the relationships and the general ideals of the group would be clearly stated. However, such agreements are normally statements of desire and are neither specific or definite.

Contracts should still be favoured in addition to partnering charters. In simplistic terms, contracts delineate what each party must do, what each party receives, time for performance, and sometimes, consequences of failure. More importantly, contracts make clear where risk is to fall. The importance of retaining contractual certainties remains constant in any relationship, although in partnering arrangements the contract becomes less the focus of attention with more emphasis on achieving a successful end result.

Ideally agreements should be used that combine a modern contract with a partnering agreement. For agreements between the main contractor and the client two contracts exist that favour partnering-based ideals. The New Engineering Contract (NEC) is a suite of contracts, including defined agreements between the main contractor and specialist contracts. The PPC2000 (ACA Standard Form of Contract for Project Partnering) is a special contract for partnering, but as yet has no standard agreement between the main contractor and the specialist.

Partnering is not a panacea for the construction industry's current, imperfect state. Even where partners endorse the critical success factors and have in place the key mechanisms to enable a successful project outcome, partnering is not an easy option. Partnering requires a real commitment, both in time, effort and trust from all partnering stakeholders. Current thinking, both from the UK and overseas, shows partnering to be a realistic challenge to traditional procurement patterns and beneficial as a mechanism for alternative dispute resolution. Partnering, however, does not guarantee that there will be no disputes or that everyone in the partnership will achieve a win-win outcome. Strategic partnering is perceived to be more beneficial to partners than project partnering, but there is a place for both types of partnering, depending upon the nature of the client and project. Current philosophy shows that partnering arrangements do not obviate the need for a contract but once the roles, risks and responsibilities are defined, the contract becomes "psychologically" less important.

....

Case Study

The Clients Charter

Following on from the recommendations made within *Rethinking Construction*, the Construction Clients Forum (now the Construction Clients Confederation), have put forward a Pact with the industry. The Pact outlines under the general heading of *Rethinking Construction's* recommendations the actions and commitments the Forum will work to in improving the industry and sets out what the Forum expects in return from the industry.

Overall the Pact commits itself to:
Clients represented on the Construction Clients' Forum commit themselves to:

- set clearly defined and quantified objectives for each project and realistic targets of achieving it;
- pool information about construction and benchmark performance across the industry;
- communicate decisions quickly through the project sponsor;
- promote relationships based on teamwork and trust, and work jointly with all our partners to reduce costs;
- where unanticipated savings in project costs result from innovative thinking, in appropriate circumstances these will be shared with the relevant parties;
- share experiences and information with the industry so that we can jointly learn to undertake improved construction;
- appraise whole-life costs, not just the "bottom line";
- use client influence to improve statutory regulation where this is burdensome;
- support training and the improvements of standards;
- educate their own decision-makers in good clientship;
- not unfairly exploit their purchasing power but look to form lasting relationships with the supply side;
- apportion risk sensibly in project contracts;
- improve their own management techniques and become better informed about the construction process.

Clients represented on the CCF would like to see specific improvements delivered in the following areas where the clients and the construction industry can work together to deliver change:

- presenting clients with objective and appropriate advice on the options and choices to meet their needs;
- introducing a "right first time" culture with the projects finished on time and to budget;
- eliminating waste, streamline processes and work towards continuous improvement;
- working towards standardisation in components where this provides efficiency gains;
- using a properly trained and certified workforce and keeping skills up to date;
- improving management of supply chains;
- keeping abreast of changing technology by innovation and investment in R&D.

Educated and experienced clients have written the Clients Charter. Its stated objectives and purposes are in line with current thoughts of procurement and general business management. Therefore, regardless of the project or nature of the client, the Charter should be used as a benchmark during the development of an appropriate procurement strategy.

....

REFERENCES

1. Bennet, J. and Jayes, S. (1995) *The Seven Pillars of Partnering*, Reading, Reading Construction Forum.
2. Bennet, J. and Jayes, S. (1995) *Trusting The Team*, Reading, Reading Construction Forum.
3. Construction Clients Forum (1998) *Constructing Improvement*, London Construction Clients Forum.
4. Egan, J. (1998) *Rethinking Construction*, London, Department of the Environment, Transport and the Regions.
5. Kelly, R. (1998) *Star Performer*, London, The Orion Publishing Group.
6. Lam, K.C., Gibb, A.G.F. Sher, W.D. (1997) *Quality Building Services - Co-ordination from Brief to Occupation*, Conference Proceedings, CIBSE Virtual Conference, London.
7. Pasquire, C. (1994) *Early Incorporation of Specialist M&E Design Capability*, Conference Proceedings, East Meets West, Hong Kong, December 4-7 1994.
8. Price, G.M. and Gibb, A.G.F. (1996) *Management of Specialist Contractor Design for Mechanical and Electrical Works, The Organization and Management of Construction: shaping theory and practice*, Vol. 2, pages 492-501.
9. Rowlinson, S. (1998) A Definition of Procurement Systems, in *Procurement Systems: a guide to best practice*, E&FN Spon, London.
10. Sheath, D.M., Jaggar, D. and Hibberd, P. (1994) *Construction Procurement Criteria; a multi-national study of the major influencing factors*, Conference Proceedings, East Meets West, Hong Kong, December 4-7 1994.
11. Shoesmith, D.R. (1995) *Whither Nominated Subcontracts for Building Services*, Conference Proceedings; Changing Roles of Contractors in Asia

Pacific Rim, Hong Kong.

12. Shoesmith, D.R. (1996) *A Study of the Management and Procurement of Building Services Work*, Construction Management and Economics, Vol. 96, pages 93-101.
13. Swaffield, L. M. and Pasquire, C. L. (1997) *Defining the Quality of M&E Services During the Early Design Stages: A Value Engineering Approach.* Conference Proceedings CIBSE Virtual Conference, London

Chapter Eleven

Engaging the Supply Chain

11.1 SUPPLY CHAIN MANAGEMENT

11.1.1 The Concept of an Integrated Supply Chain

Integrating separate parties to form a team that is dedicated to delivering a mutual project consists of two distinct operations, supply chain management and logistics. Each is a distinct process, but both interact to form a cohesive delivery package.

Buildings are not unique in their requirement for a multitude of people to deliver a cohesive package. Most other industries work on the basis of using sub-manufacturers and component suppliers and face similar problems to those of construction - continuity of supply, appropriate and timely information, expediency of delivery and control of quality. Supply chain management sets out the *relationships* between the parties involved, whilst logistics determines the *procedures* within the relationship.

The package must be delivered by the diversity of people involved, each of whom has a unique role and influence on the project:

- the clients and end users who are served by the building services industry and who have an influence on it;
- the designers who design and integrate the building and the building services;
- the manufacturers/suppliers who design, produce and supply the components and equipment that constitute the service systems;
- the contractors who translate the designer's concepts and innovations into reality and who also plan, co-ordinate and install the services;
- the facility manager who maintains the building.

It is clear that for best project results, a team approach is needed to carry out all these interrelated activities. They have to be undertaken by each of these five parties as an integrating unit since:

- the client's contribution can influence the services designer's design and other subsequent activities;
- the designer's design can influence the physical installation of building services and maintenance aspects;

- the contractor's working and expert advice can affect the decision-making process of the design and management of the installation;
- services design and contractor's decisions on practicality, constructability and cost can influence the manufacture of M&E services, products or components;
- service's design and installation can have a significant impact on the management of the maintenance of M&E services, and thus the usefulness of a building.

To first understand the nature of supply chain management, it must be understood that the contribution made by each party supplying to a project is valued separately. Simplistically, if the two most important aspects a project are risk of the system working and cost of the system, a simple matrix can be developed.

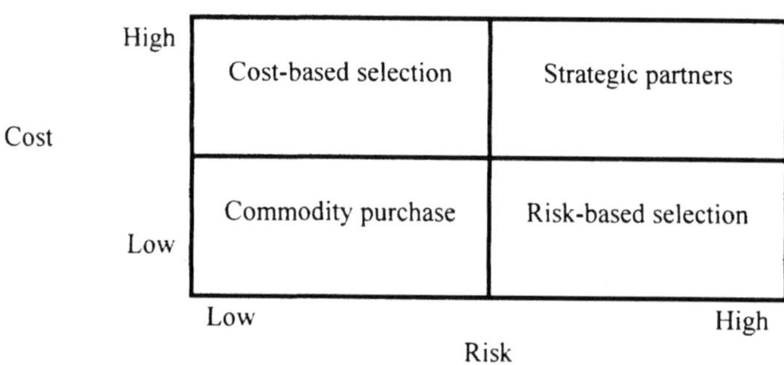

Figure 11.1 Supply Decision Matrix

Organisations supplying high-risk / high-cost items, such as boilers to a hospital, are critical to the success of the installation and therefore should be dealt with as a strategic partner, selected on a weighted quality assessment. In contrast low-cost / low-risk suppliers have little influence on the project. Therefore, price becomes the dominant deciding factor in their selection.

Using a simple assessment matrix like that above immediately categorises each supplier as to their role in the supply chain and any weighting that should be applied.

....

Case Study

Using a similar system to that above, BAA on Project Genesis, their landmark project using supply chain management, organised each work package into three key groups:

- **key packages:** those with high importance top project success, typically cladding, building services, lifts and foundations.
- **secondary packages:** those with average importance to the project, typically landscaping, metalwork and decorations.
- **organisational packages:** those that provide support to the project, but do not add to the physical project, e.g. hoarding, security and site accommodation.

Each work package was then analysed and a suitable strategy was developed using one of three sourcing strategies:

- **supply chain analysis:** where added value could be developed;
- **package bid:** where competition was desired, either commercially or some other criteria;
- **single source strategy:** for packages where a long-term partner is used or one supplier having superior abilities or service.

Each package was finally analysed as to whether designer - or contractor - lead design was preferable.

....

Beyond this assessment, supply chain management is about managing the operations between each company that forms the chain of supply for a particular system. It is particularly important for building services due to the number of components involved, the diversity of design requirements and the global spread of suppliers. To work, each supplier must work on a common strategy, underpinned by common systems.

The success of supply chain management in other industries has been attributed to a multitude of information technologies and management strategies, most of which are undeveloped for use in construction. The disruptive and erratic nature of demand does not allow for the formation of such strategies on a single construction project. However, given the speed of information technology development and the importance of delivery logistics it is expected that developments in this area will rapidly reach construction.

What can be used are the general lessons learned from other industries, which can be categorised into three key areas of improvement.

1. **Information:** information should be freely shared using compatible software. The use of project websites and project intranets have begun this move. By freely sharing information electronically the state of progress of any work, either design or construction, can be viewed by all parties' members. This allows greater anticipation of requirements and eliminates endless paper chases for correct information.
2. **Process alignment**: going beyond the simple provision of information, a considerable amount of waste is generated by transferring information and suppliers from one company's systems to another. Process alignment works toward ensuring each party's processes are compatible, allowing the direct and seamless transfer of goods and information. Furthermore, work programmes are harmonised allowing for quicker transfer as no waiting is needed between operations.
3. **Operation efficiency:** commonly referred to as lean construction or management, operation efficiency has two goals – to eliminate as much waste as possible and ensure that each step in the process contributes to the overall quality. Much of this is to do with reducing any operation to a minimal basis, ensuring that it is simple and thereby eliminating errors.

....

Case Study

Rethinking Construction
Prepared by the UK Government in response to clients' concerns with the construction industry, the report highlights a number of improvements that can be made. Bringing the supply chain together is expected to address the following issues, most of which are imperative to a quality building services installation. The report states:

- **Suppliers and subcontractors** have to be fully involved in the design team. In manufacturing industry, the concept "design for manufacture" is a vital part of delivery efficiency and quality, and construction needs to develop an equivalent concept of "design for construction".
- The **experience of completed projects** must be fed into the next one. With some exceptions the industry has little experience in this area. There are significant gains to be made from understanding client satisfaction and capturing technical information, such as the effectiveness of control systems or the durability of components.
- **Quality** must be fundamental to the design process. Defects and snagging need to be designed out on the computer before work begins on site. "Right First Time" means designing buildings and their components so that they cannot be wrong.
- **Designers** should work in close collaboration with other participants in the project process. They must understand more clearly how components are manufactured and assembled, and how their creative and analytical skills can be used to best effect in the process as a whole. There is no longer a place for regime of design fees based on percentage of the cost of a project, which offers little incentive to build effectively.
- **Design** needs to encompass whole life costs, including energy consumption and maintenance costs. Sustainability is equally important. Increasingly, clients take the view that construction should be designed and costed as a total package including costs-in-use and final decommissioning.
- **Clients** too must accept their responsibilities for effective design. Too often they are impatient to get their project on site the day after planning consent is obtained. The industry must help clients to understand the need for resources to be concentrated up-front on projects if greater efficiency and quality is to be delivered.

Source: Rethinking Construction, DETR

11.2 INTEGRATION AND CO-ORDINATION

11.2.1 Roles and Responsibilities

Due to its dynamic nature and inter-operability, building services must be fully integrated and co-ordinated for success. Success of co-ordination, and therefore the project, is related to appropriateness of a particular procurement system. Although integration and co-ordination are affected by the strategic decision, their actual implementation is controlled by the mechanisms enabled during the project. It is the role of procurement only to ensure that compatible parties are selected and the overall strategy developed enables the mechanisms to be developed.

The detailed nature of integration and co-ordination has been debated elsewhere within the text. What must be considered during the selection of project partners are key attributes of the organisation that will allow success to be achieved. For contracting and engineering-based organisations both must have compatibility in:

- information systems
- quality systems
- managerial frameworks
- business nature
- management paradigms
- project approach
- understanding of technology.

All projects will involve a multitude of parties. The importance of good co-ordination has been expedited by an increase in subcontracting of specialist works. As technology expands in complexity and breadth of systems, the specialist engineering industry also fragments and becomes further specialised by both type and number of organisations. The primary mechanical and electrical contractors mainly act now as main contractors with the role of co-ordination becoming their dominate role. Typical work packages sublet by a building services contractor include:

Mechanical

- ducting
- insulation
- controls
- commissioning
- O&M manuals/record drawings

Electrical

- lightening protection
- fire alarms
- security

Public Health

- chlorination

Specialist Services

- BMS/controls
- substation
- high-voltage switch gear
- security
- data
- telecoms
- fire protection / alarms
- generators / UPS
- kitchens and cold rooms
- process/medical gases
- commissioning.

To implement an effective procurement strategy that recognises integration and co-ordination certain key steps must be inherent within the process. It is fair to say that most aspects of good integration and co-ordination are dependent upon the specific systems adopted during construction. But procurement must play its part and establish the correct principles from the outset.

- **Step 1.** On the establishment of the overall concept design for the project the list of technical systems and requirements is agreed with the client and then a preliminary work breakdown structure needs to be agreed. This must be based upon the intended procurement strategy, which must respect the responsibility for control and the level of breakdown in which the packages are to be procured.

- **Step 2.** Develop a procurement strategy based on these work packages, with an assessment of project objectives.

- **Step 3.** A design responsibility matrix must be completed, clearly stating who is to design what, who checks it and where the information is to be passed on to. This must be made in conjunction with a design information matrix showing the flow of information between the respective parties.

- **Step 4.** Feed the information obtained in Step 3 into the contract documents before tendering or other selection procedures are used. This ensures from the outset that all parties understand their obligation for successful integration.

- **Step 5.** Agree programmes for design, procurement and construction, respecting the amount of work required to be undertaken by each party. Allowance is to be made for checking, revisions and co-ordination of other information. The preparation and submission of shop drawings must include the correct sequence for each trade.

- **Step 6.** Assess each party's general corporate approach for compatibility against the project.

- **Step 7.** Once parties are all on board hold a separate series of co-ordination meetings that include both technical considerations and sufficient personnel motivation for a team.

11.2.2 Work Breakdown Structures

Properly structuring the building services installation, in terms of organisational responsibility and delineation between items of work, is imperative for both successful integration and the implementation of procurement strategy. Structuring a project not only provides a framework for co-ordinating the works but assists in establishing suitable mechanisms for its effective control.

Structuring a project defines the first step in developing a procurement strategy. Deciding how the work should be broken down determines the number of individual parties to be procured, the amount of co-ordination needed, the number of organisational interfaces and the extent to which the procurement strategy can affect the entire supply chain. The defined structure then by default determines the people involved and on which level they must communicate.

The intersection of these two structures – organisational responsibility and delineation of tasks - identifies the work scope for each package. Furthermore, as each package links to form the whole, the project's objectives can be broken down into goals for each package with appropriate benchmarks for cost, time and quality.

Structuring

Work breakdown structures are simply family tree divisions of a system. Each has a defined subdivision of work with clear deliverables and boundaries of operation. These divisions can then be overlaid with subdivisions of other categories, such as cost budget, time allocations or management hierarchies, to establish a fully integrated three-dimensional model of the project.

To be successful the project must be broken down logically and objectively. With building services the final level to which the project is broken down will vary between branches, depending upon the technology involved and extent of work. The work package must be within the capability of a single company and therefore some reference to the market place and possible organisations must be made.

There is no limit to the number of levels the work should be broken down into, but they must be compatible with the procurement strategy adopted.

A typical services installation can be broken down as follows:

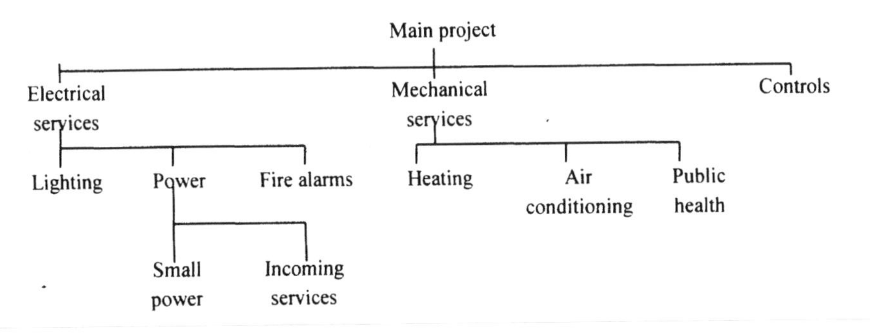

Figure 11.2 Project Work Breakdown Structure

The project structure will normally be determined by the main project structure. The same structure is used for cost coding, accounting and quality systems structuring.

For procurement, the same chart could be further refined to reflect actual parties and work packages. Mechanical services are illustrated in Figure 11.3.

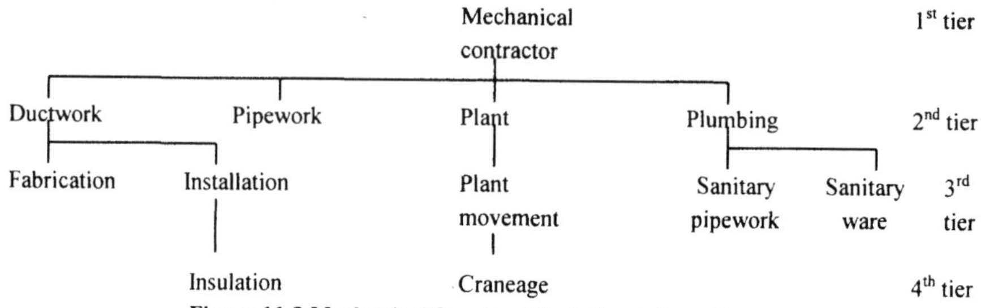

Figure 11.3 Mechanical Services Work Breakdown Structure

Although this illustrated breakdown is far from a definitive guide (or even a logical one!), it is useful to illustrate the following points:

- **mechanical contractor:** works as a first-tier contractor, procured either through the main contractor or as part of the primary project team. All other mechanical services are procured directly through this contractor.
- **ductwork:** may be managed as a second tier directly by the mechanical contractor, but split between fabrication and installation (becoming third-tier suppliers). The insulator of the ductwork may then act as subcontractor to the installation contractor (acting as a fourth-tier supplier).
- **plant:** is procured directly by the mechanical contractor and becomes freely issued to the other contractor, except the major plant which is installed by a third-tier plant movement contractor.
- **plumbing:** may be procured as a direct package to a second-tier subcontractor who subsequently subcontracts out the pipework and sanitary ware installation.

There are key rules that should be followed when structuring the project. Guidelines include the following:

- The subdivision of a work package must be worthwhile for management, control or commercial reasons. It must have a defined element of work with specific deliverables.
- There is no necessity for work packages to be broken down into the same number of tiers.
- Each package must be broken down into logical parts.
- A cost breakdown structure must mirror the work breakdown.
- Each work package must be capable of being procured as a single entity, with respect to industry norms, capability and capacity.
- The extent of work package breakdown must be inline with the procurement strategy. The controlling first-tier party must be capable of managing the break down.

....

Case Study

Current Problems Facing Specialist Engineering Contractors
In 1999 Reading Construction Forum published a report entitled *Unlocking Specialist Potential.* The report studied the current problems facing specialist contractors and proposed a number of solutions. These were mainly based around specialist contractors participating within the project design development and providing services on a product basis.

The problems highlighted within the report can be used as a checklist during the procurement strategy development to ensure it has respected and addressed the issues. The report stated current problems include:

- unrealistic project programmes
- undue emphasis on cost rather than value, producing fierce competition and reduced margins
- dutch auctions
- perceived poor status of specialist contractors
- onerous contract conditions and inappropriate unloading of risk
- unreasonable payment procedures
- lack of understanding of the risks involved and their consequences
- unclear statements of requirements and unambiguous project information and tender packages
- insufficient focus on customer-supplier relationships
- lack of clarity of agreed requirements with customers and suppliers
- irregular flow of work
- insufficient project planning resulting in overlapping of functions and waste
- poor flow of information
- underdeveloped management and interpersonal skills.

11.3 SELECTING CONTRACTORS AND CONSULTANTS

"...We were told, and we saw for ourselves, that problems frequently occur on the later stages of a project, on the mechanical and electrical services and on fitting out. More time and effort spent on getting that right should clearly repay itself many times over. Departments should give early attention on building projects to putting the right team in place to ensure proper planning, design and co-ordination of the m&e services."

Source: Construction Procurement: An Efficiency Unit Scrutiny, HM Treasury.

11.3.1 Assessing Competence and Ability

Assessing possible members that are collectively to form the building services team is the most critical function to be executed from the procurement strategy. The appointment of inappropriate contractors or consultants will diminish the effectiveness of the strategy and may compromise the overall quality of the end product. To some it may appear to be inappropriate to discuss the selection of consultants and contractors collectively. But to form a cohesive team core abilities and attitudes must be compatible with all parties, thus dictating a common set of measures.

Once an organisation has been assessed using the common set of criteria, other criteria can be developed and used that are specific to their role. These will not only vary from contractor to consultant, but will vary between individual roles. For example, the specific criteria used to assess the lead designer will be different to that used for a specialist building controls consultant. The specific criteria used must always be set within the context of the project and that of the general industry. Criteria such as profitability must be compared against the norm for the sector, not against a general business measure.

Appraising and selecting appropriate parties is based on the following process:

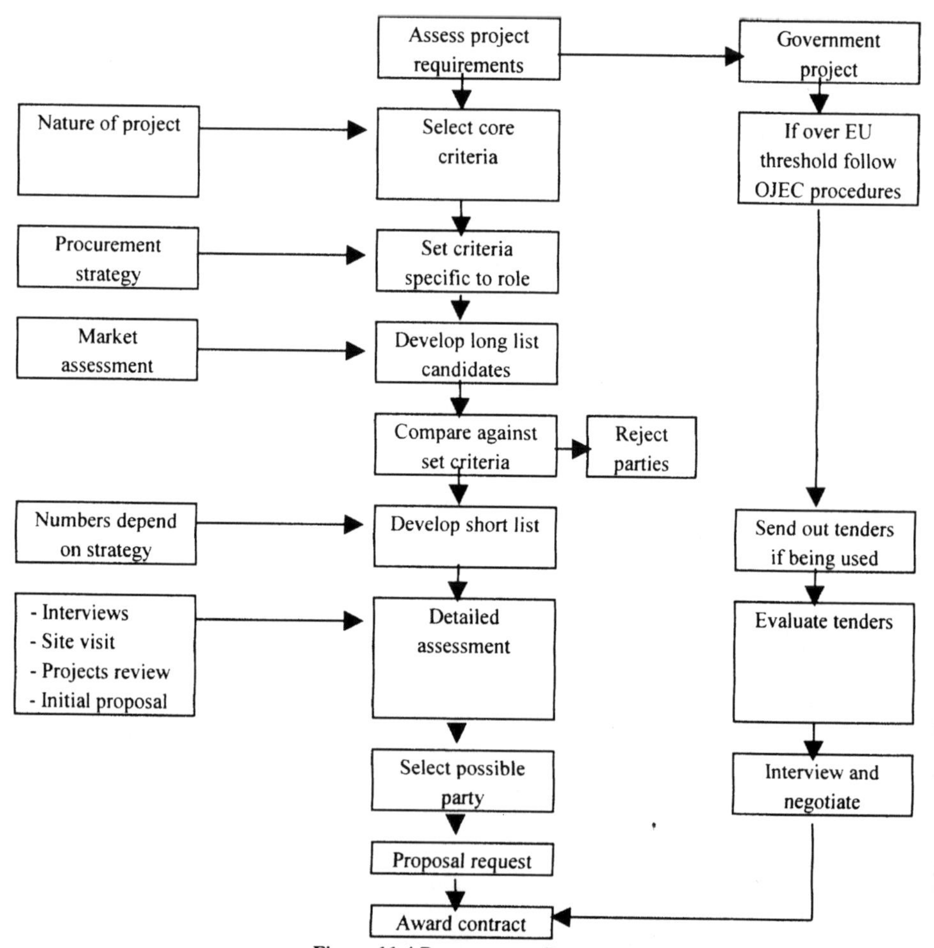

Figure 11.4 Procurement Process

The core requirements used for the selection of suitable designers and contractors will vary according to:

- the nature of the project
- the size of the project
- the procurement strategy for both the services and main project
- contract conditions
- the location of the project
- the nature of the work to be executed under the appointment.

Appraising all members of the building services team is particularly important when considering contracts that usually lead to long-term agreements. Such contracts require the parties to have a close, flexible and continuing good relationship. Many large organisations usually maintain a list of approved suppliers from which an adhoc list to suit the requirements of the job may be produced.

Pre-qualification

Regardless of the final procurement strategy agreed, or whether the possible contractors and consultants are known from the outset, some form of pre-qualification will exist. If the parties have worked together before on a continual basis, this may be as simple as a phone call or letter of invitation to join the project. Normally it would involve some form of assessment.

Although using known contractors and designers is desirable from the point of view of offering comfort and a minimal learning curve, it excludes new companies from bringing in fresh ideas, or instilling sufficient drive to ensure efficiency is maintained.

Prequalification should not be seen as an overly competitive exercise, but more of an exploration from both sides as to whether there is sufficient desire on both parts to work together. From the clients side the basic purpose of carrying out an assessment is to ensure that the party is capable, suitable and competent to carry out the work and to confirm that they are willing to either become part of the building services team or tender for the works. From the contractor's or consultant's point of view, preselection allows them to assess whether the project is desirable, both commercially and managerially, whether the client is financially sound, and how the relationship they will have with the other parties will work: in short they want to know what the procurement strategy is.

The overall nature of the project, whether a single project or a multiphase or serial project, will affect the emphasis of the criteria selected. In the case of project contracts the client is most likely to be primarily concerned about the contractor's competence to undertake the project in terms of cost, time and quality. In the case of a project that is going to be extended over a long time period, then more esoteric criteria such as people's attitudes, commitment to quality improvement, information technology levels and corporate stability are more likely to dominate.

Assessment Criteria

The establishment of suitable assessment criteria forms the basis of the decision for the suitability for an organisation's selection. As no party should ever be procured on a money only basis, criteria will be inputted into the evaluation that will be used to calculate the value for money of the commercial proposal.

Assessment criteria are divided into two distinct groups. Core criteria are based on the project and procurement strategy. All parties that will form the building services team must meet these criteria in order for the requisite teamworking and communication to be developed. As a minimum they should include:

- experience with teamworking
- experience with the management style to be used on project
- experience with the technologies involved
- experience with the type of project
- experience with other team members (desirable but not compulsory).

Organisations not fulfilling the above criteria should not be assessed any further. The inability to meet any of the above will cause detriment through the entire building services team.

The second set of criteria is specific to the organisation and technology involved. Its assessment is best done through a series of questions, some of which may or may not be relevant to the project. The questions posed under the following headings are suitable for both consultants and contractors.

Track Record: The track record of a company should be an overall assessment of the company's work record. It should scrutinise the number of projects completed, based on an assessment of the following for each project over the past 5 years:

- programmed completion vs. actual completion
- final account cost vs. tender amount
- client type
- repeat work from the same client
- the role in the project
- services provided.

What is being looked for are general patterns of service quality and general experience with similar projects. If the vast majority of projects undertaken by the contractor or consultant end up over budget, late, and with the client giving no further work, then sufficient danger signs are present to exclude that organisation from further consideration. On a similar basis, if the organisation being considered is unable to present the above statistics then this too is a danger sign.

Previous Experience: Beyond the track record it is ideal to find an organisation that has direct experience with the specific type of project proposed. If the project is a new build surgical unit, the track record may show general experience with hospitals, but what may be being sought is experience with surgical units specifically. It is worth making this distinction as few organisations may have the experience and the best possible alternative may be sought.

Assessment of previous experience should include detailed analysis of performance on the contract including contacting the client. A questionnaire-based survey may be particularly useful.

Resources: Resources include all the plant, equipment facilities and general administrative staff required to complete the works. Although this type of assessment is normally reserved for the contractor, a significant number of contracts fail due to lack of resources available with the designer. Design resources can include levels of information technology, the number of CAD operators, specific engineering types etc. To make a fair assessment will require some analysis beforehand to compare the anticipated resources with those able to be provided by the organisation. Furthermore, it will have to be assessed with other projects currently being handled.

Management: Senior managers are the drivers of a company. They will also make the strategic decisions regarding a contract. Therefore, they should be in tune with the project needs and have sufficient capacity to ensure that the company's resources, both staff and otherwise, are committed to the project. Assessment could include their personnel experience, length of time with the company, qualifications and general duties.

Capacity: Factors to be included in the evaluation of a organisation's capacity include the number, experience and qualifications of professional and technical staff responsible for providing the service, the amount of work in hand, the availability of resources, the use and management of subcontractors.

Tradespeople and Support Staff: Under-resourcing of labour on site often leads to delays and quality problems. Similar to that of capacity, detailed assessment of site staff should include not only the quantity, but also their location, skill level, training and current deployment on other projects.

For consultants this will include designers, engineers, CAD operators and other necessary support staff.

Quality: It is not enough to ask whether the company has quality procedures certified under the BS EN IS0 9000 or an equivalent standard. What must be checked is that the certification covers the services relevant to the project and moreover, that the quality system is actually being implemented and improved.

Project Management: It must be checked that the project management methodology is well documented, practised and adhered to effectively. The project manger will often dictate the success of a project. Detailed assessment must be made of the actual project manner proposed for the project. Issues such as qualifications, experience, dedication to the project (is it their sole project or will they be looking after others?) and general attitude, need be assessed.

Supply Chain Strategy: Having long-term strategic relationships with the supply chain is imperative for project success. Contractors should have established relationships with core suppliers and contractors. Proof should be asked for that the relationship is a formal one based on a common set of goals and the implementation of a properly considered supply chain management strategy. This could include working towards common e-procurement and information technology systems.

Risk Management: Transferring risk to another party is only successful if that party is capable of handling the risk. Assessing whether an organisation has a risk management procedure in place ensures the company is capable of handling risk and, moreover, understands the serious nature of the subject.

Financial Assessment: Requesting information covering the last three years is important because trends must be carefully examined. Care should be taken if doubts exist about a supplier's financial standing, particularly when a large value, long duration contract is being let. The client must be confident of a supplier's ability to fulfil the contract.

Financial warning signals may include:

- falling profitability
- increasing debtor days
- increasing debts and creditor days
- increasing stocks, slower stock turnover
- deteriorating liquidity
- over-reliance on short-term debt
- high gearing
- late production of accounts
- qualified accounts
- changing auditor's and/or banker's name
- cash draining from the business
- major reductions in staffing.

Site Visits

Supplier and reference site visits can be undertaken in parallel with general assessment. Visiting a company's premises and current projects allows assessment as to whether the information supplied regarding the above is actually being carried out and gives a true reflection of the company.

For equipment suppliers or specialist engineering contractors who are to supply a high proportion of fabricated goods, then the manufacturing premises should also be visited. The procurement team should carry out a similar assessment to that above to validate a supplier's financial, commercial and technical credentials. Visits should be carefully planned and the coverage might include a review of management and technical competence, quality control procedures, the approach to training, and the current order book.

Visits can also be made to current or previous projects of a potential supplier, both in the public and private sectors. The purpose is to get customers' views of the supplier and its services, and thus a more objective assessment of its capability and suitability.

Suppliers tend to select as referees, customers who would give a favourable report, so they may not be entirely objective. A better approach is for the evaluation team to ask for details of all relevant contracts won in the last 2-3 years and then to make their own selection. The customers selected must of course be willing to co-operate in the provision of feedback.

Questions to ask such customers include:

- were they satisfied with the supplier's performance in providing the service required?
- has the supplier the ability to work effectively with the client organisation and culture?
- what were the supplier's approach and attitude to change?
- did they make any general observations regarding the supplier's strengths and weaknesses?
- were there any problems with the contract terms?
- what would the customer have done differently if they were going through the process again?
- would they use the supplier's services again? If not, why not?

Maintenance

The requirement for maintenance will immediately follow a new installation. The installation contractor may have a maintenance division, which is capable of fulfilling the maintenance task. This may be good practice, especially during the warranty period, unless the installation contract has not been entirely equitable. However, many maintenance divisions of installation contractors operate completely separately, and the performance of the maintenance division should not be pre-judged on the performance of the installation division.

If under the procurement strategy it is likely the maintenance will be carried out by the installation contractor, then a separate assessment should be undertaken of their maintenance abilities. Both assessments will then need to be considered together to ensure that a suitable contractor is available to carry out both parts of the contract.

11.3.2 CDM Requirements

The *Construction (Design and Management) Regulations 1994* (CDM) imposes particular conditions on anyone procuring a contractor or design service. Other than government work that comes under various accountability and EU Regulations, the CDM Regulations are the aspect of regulation that must be used to assess a contractor's competency.

The regulations apply to every construction project lasting more than 30 days and employing 5 or more people. The regulations require every client to appoint a Planning Supervisor and a Principal Contractor, who are competent and provided with sufficient resources to perform their functions as dictated by the regulations.

Although these regulations apply to the actual design and construction of the project, they do affect procurement in that a minimum competency threshold and certain duties are imposed by the regulations. These duties require the three principal roles to carry out the following:

Designer's Duties

- mitigate possible risks during design
- provide information on hazards
- co-ordinate with other designers over risk
- ensure the client is aware of CDM duties.

Planning Supervisor's Duties

- prepare a health and safety plan and a framework for the project
- ensure designers are considering health and safety issues
- ensure co-operation between designers
- help to pre-qualify the principal contractor (if requested)
- prepare a health and safety file.

Principal Contractor's Duties

- check, adopt and further develop the health and safety plan
- ensure co-operation and exchange of information between contractors
- ensure contractors comply with the rules
- ensure employees receive adequate information and training.

The only duties imposed on specialist engineering contractors, other than when they work as the principal contractor, are to forward relevant health and safety information to the principal contractor and comply with the project health and safety plan.

11.3.3 Weighted Assessment Matrix

To assess a company objectively requires the establishment of a weighted assessment matrix. The purpose of the matrix is to provide an objective assessment of an organisation's abilities within the context of a particular project. The assessment should include minimum standards that must be obtained, together with weighting the importance of each criterion.

The matrix works by assigning ideal scores to several key criteria. The ideal scores should be based on the project's needs and client objectives. This should be carried out in an impartial manner. Possible project partners are then assessed and scored against the stated criteria. Companies scoring below minimum thresholds should not be considered further. After this initial assessment, interviews should be held with the highest scoring parties to agree the successful candidates. The number of final candidates selected will depend upon the final method of tendering, be it competitive, partnering, two-stage etc.

This matrix can then be developed further into a quality price mechanism once tenders have been received. Working in the same manner, the matrix has the tendered costs assigned to it, with appropriate weighting between the quality criteria and price. The type of project decides the weighting between quality and price.

The *Office for Government Commerce Procurement Guidance Note 3 – Appointment of Consultants and Contractors* recommends the following weightings for both quality criteria and quality / price ratios (Figure 11.5):

Type of Project	Indicative Quality / Price Ratio	
	For Contractors	**For Consultants**
Feasibility studies and investigations	n/a	80/20 to 90/10
Innovative projects	20/80 to 40/60	70/30 to 85/15
Complex projects	15/85 to 35/65	60/40 to 80/20
Straightforward projects	10/90 to 25/75	30/70 to 60/40
Serial projects	5/95 to 10/90	10/90 to 30/70

Quality Criteria	Suggested Weighting Range
Practice or company	20 – 30%
Project organisation	15 – 25%
Key project personnel	30 – 40%
Project execution	20 – 30%

Figure 11.5 Quality / Price Ratio

Weighted assessment matrixes are not foolproof. They rely on objectively assessing the information provided by the contractor. Whenever possible all information should be verified through company visits, interviews and assessment of detailed documentation.

REFERENCES

1. Barton, P.K. (1976) *Co-ordinating Services Sub-contractors: a systems approach,* Building Technology and Management, October 1976.
2. Egan, J. (1998) *Rethinking Construction*, London, DETR Publications.
3. HMSO (1994) *Construction (Design and Management) Regulations 1994*, London, HMSO Publications.
4. HM Treasury (1998) *Procurement Guidance Note 3 – Appointment of Consultants and Contractors*, London, HM Treasury Publications.
5. Levene, P. (1995) *Construction Procurement By Government: An Efficiency Unit Scrutiny*, London, HM Treasury.
6. Saad, M. and Jones, M. (1998) *Unlocking Specialist Potential*, Reading, Reading Construction Forum.
7. Schal (1997) *Recommendations for Best Practice*, London, Schal in-house publication.

8. Slack, N. Chambers, S. and Johnston, R. (2001) *Operations Management*, London, Pearson Education Limited.
9. Wild, J. (1997) *Site Management of Building Services Contractors*, London, E&FN Spon.

Chapter Twelve

Performance Modalities

12.1 PERFORMANCE MODALITIES

This chapter develops the third set of modalities and concentrates on the human and performance elements of the project. The three topics covered are: partnering, supply chain management and key performance indicators.

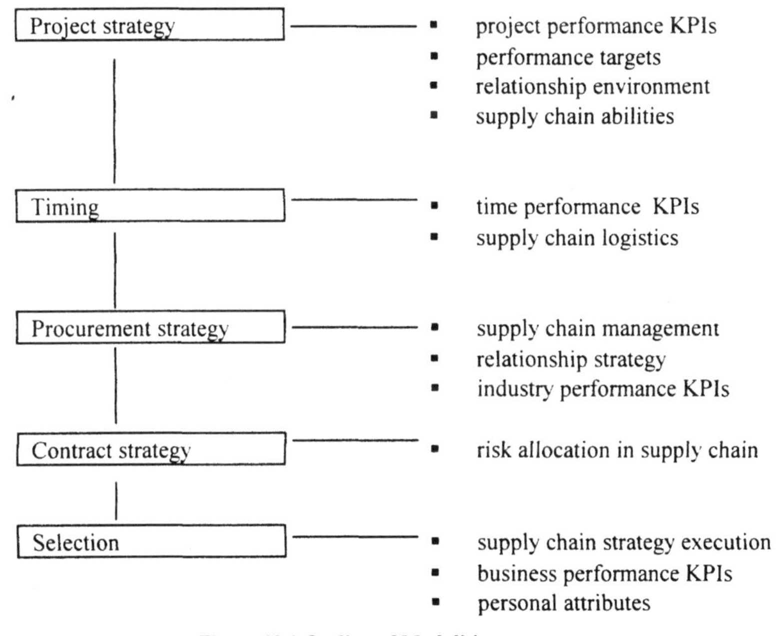

Figure 12.1 Outline of Modalities

These three modalities do appear as separate items within the decisions chain, as their basic nature demands consideration at all phases of the project. Having stated that, at certain times each will play a significant part at particular stages of a project. Supply chain management is normally considered at both the initial decision of work package formation and the selection of particular organisations to execute the work. Partnering, and its inherent decision over the types of people and the relationship they will have with each other, has no particular start or finish, as it is an underpinning of the overall procurement strategy. Although obviously the true execution of the partnering philosophy can only be undertaken once all members of the building services team are on board.

Finally, key performance indicators can be used in a variety of ways depending upon the particular stage of the project. They can be used either to set the performance targets for the project or to assess the credibility of team members currently under consideration. Unfortunately, as yet, no comparable indicators exist for consultants, but reference can be made to the various independent market research works undertaken by such organisations as BSRIA to form similar performance measures.

12.2 SUPPLY CHAIN MODALITIES

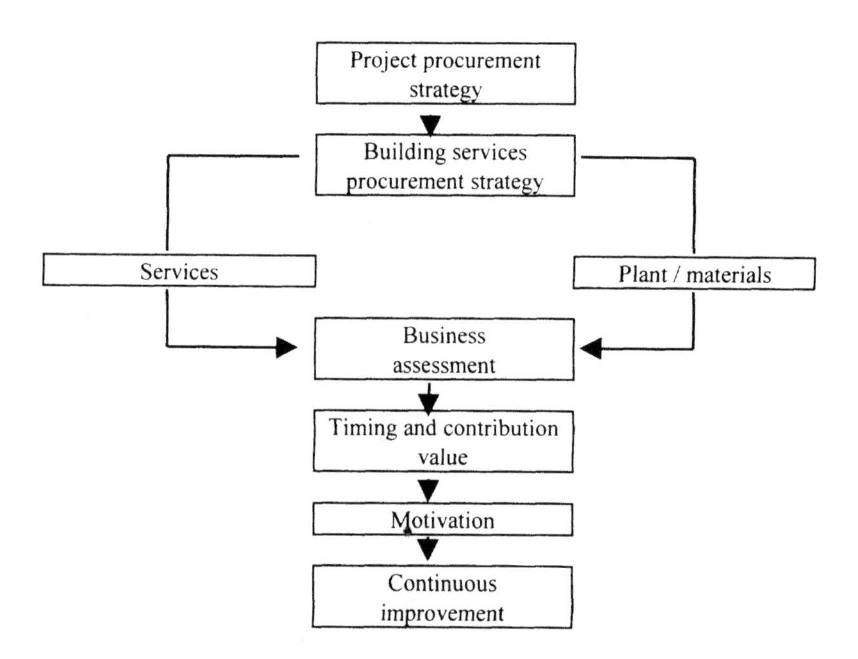

Figure 12.2 Outline of Supply Chain Decision Process

The procurement strategy for the project will dictate the relationships between the end client and the building services supply chain. To form a cohesive relationship, a two way decision must be entered into: the procurement strategy will dictate the manner in which services and plant/materials are procured but equally the supply chain must be capable of meeting the strategy adopted.

Therefore, the modalities provided are to analyse this relationship. The procurement strategy adopted for the building services must first analyse the market that is being considered, by analysing the general characteristics of the industry. Considerations include those listed in Table 12.1:

Supply Market	
Area of Consideration	**Considerations**
▪ Market size	– is there a ready supply of competitors? – does the market have capacity capability?
▪ Level of competition	– is detailed cost accountability required? – is competition needed to increase performance?
▪ Legal considerations	– is project subject to legal procedures e.g. EU? – is market restricted by legal requirements?
▪ Technological requirements	– who is driver of technology: client or market? – is the market used to the technology involved?
▪ Industrial relations	– will unions be involved? – are labour relations currently stable?
▪ General economic health of market	– is the market rising or falling? – what is the perceived general market health?
▪ Inflation rate	– does market inflation match rest of economy? – are market prices stable?
▪ Source of supply (for materials)	– are materials dependent on overseas markets? – are extended delivery times permissible?
▪ Purchasing strength of client	– does client have useable purchasing strength?

Table 12.1 Supply Market Considerations

From assessing the answers to the above a general market profile should be developed that will determine the general boundaries for decision making. Obviously any procurement strategy developed needs to be within the limits determined.

Supplier Considerations	
1. Quality of relationship	– is relationship established or new?
2. Type of relationship	– to be contractual or informal?
3. Level of trust	– does organisation appear trustworthy?
4. Performance record	– undertaken similar work before?
5. Production technology	– is supply dependent on manufacturing?
6. Technology supplied	– do they have previous experience with it?
7. Head office facilities	– is full administration support available?
8. Level of flexibility	– does organisation have rigid systems?
9. Quality record and systems	– is it documented and audible?
10. Financial stability	– full accounts available?
11. Head office location	– local, regional or international?
12. Project value proportion of total turnover	– will there be a drain on resources?

Table 12.2 Suppliers Considerations

Once determined, a matching of possible suppliers can be undertaken. As with the market, a review using the considerations shown in Table 12.2 must be undertaken to match the project requirements with capable organisations.

Selection Assessment				
Project *A New Building* Organisation *Any Company Limited*				
Personnel Criteria	Threshold score	Criteria weighting %	Score awarded	Weighted score
Qualification	*50*	*25*	*80*	*20*
Management ability	*60*	*25*	*75*	*19*
Experience	*60*	*25*	*60*	*15*
Attitude	*50*	*25*	*80*	*20*
		100		
			total A	*74*
Financial Criteria	Threshold score	Criteria weighting %	Score awarded	Weighted score
Profit and loss	*60*	*20*	*65*	*13*
Retained capital	*40*	*20*	*70*	*14*
Insurance provision	*50*	*30*	*70*	*21*
Work in hand	*60*	*30*	*60*	*18*
		100		
			total B	*66*
Technical Criteria	Threshold score	Criteria weighting %	Score awarded	Weighted score
Previous experience	*60*	*30*	*65*	*20*
Resource availability	*50*	*30*	*50*	*15*
Specialist experience	*40*	*5*	*40*	*2*
Quality assurance	*50*	*3*	*55*	*17*
IT resources	*50*	*5*	*55*	*3*
		100	total C	*57*

Total Score

	Weighting %	Score	Total
A	*30*	*74*	*22*
B	*30*	*66*	*20*
C	*40*	*57*	*23*
Total Assessed Score			***65***

Notes:

- number of criteria variable, but should be no more than 10
- criteria weighting should equal 100% per section
- all factors and criteria stated in italics are variable
- assessment is suitable for any type of service or supplier
- threshold score is minimum score acceptable

Figure 12.3 Contribution and Value Assessment

The answers to questions posed in Table 12.2 can be answered in relation to the project objectives, using an objective marking system, as that provided in Figure 12.3

The initial procurement strategy should be viewed as to how it can be improved upon or further developed with additional value from the supply chain. Beyond their general competence to undertake the work, consideration must be given as to a company's ability to provide added value through their corporate structure and available resources.

Supply Chain Added Value Assessment	
Key Factor	**Consideration**
Long-term contribution	▪ Do organisations have a continuous improvement policy? ▪ How are lessons learned going to be recorded? ▪ Do suppliers fully understand project needs? ▪ Will installing organisations be undertaking maintenance contracts? ▪ Is there a reward-sharing scheme for improvements?
Integration and co-ordination	▪ Do organisations share compatible IT systems? ▪ Does a project IT strategy exist? ▪ Is concurrent engineering to be utilised? ▪ Are supplier development programmes in place?
Management of interfaces	▪ Is there a dedicated information co-ordinator? ▪ Has an information and responsibility matrix been developed? ▪ Are design responsibilities clearly defined?
Integration of suppliers	▪ Are all opportunities to combine design and construction being taken advantage of? ▪ Are work packages and division of suppliers based around technology clusters?
Minimum supply chain	▪ Have supplier numbers been reduced to a minimum? ▪ Are the supply chains between manufacturer and installation as short as possible?
Lean supply	▪ Are processes for the project standardised? ▪ Are processes between companies in the same chain standardised? ▪ Are processes as simple as possible?
Continuous improvement	▪ Do companies have compatible quality systems? ▪ Does the project have a quality system?
Logistics	▪ Can materials be co-ordinated off-site for combined palletised and tagged deliveries?
Strategic supply	▪ Have all work packages been assessed for risk and value contribution?

Table 12.3 Supply Chain Added Value Assessment

12. 3 PARTNERING

12.3.1 Using the Modalities

As previously stated, the approach to which partnering is undertaken varies through the particular stages of a project. It moves from the high level strategic consideration during the setting of the project objectives to the tactical exercises used to develop the necessary teamwork.

The decision process surrounding partnering can be viewed as having three streams:

1. **Economic:** nearly all projects have their feasibility and strategy determined by economic means. Partnering must provide some form of advantage, either as a direct cost reduction or through advantages provided by shorter times or improved quality etc. Judging whether partnering will provide economic advantage is based upon assessing the corporate goals. Between these two streams lies the principal decision that must be made.
2. **Social:** social assessment determines whether partnering enhances the project objectives, and whether the project is able to foster a sufficient environment to gain advantage from it. Starting with the corporate goals of the project, a stage assessment is posed. It at anytime the answer of no is judged appropriate for the question, then it is unlikely that the project will benefit from the added effort of undertaking partnering.
3. **Cultural:** having passed the economic and social hurdles, and assuming it desirable to undertake partnering, then actions must be undertaken that provide the necessary environment that will promote a culture of co-operation and trust.

Three modalities are provided:

1. A decision flowchart to determine the appropriateness of using partnering (Figure 12.4).
2. A set of detailed considerations that need to be incorporated into the procurement strategy and dealt with during the formation of the building services' team.
3. A list of key activities that should be undertaken at particular stages of a project (Figure 12.5).

The procurement strategy must embody the final decision and issues determined through this assessment process. Partnering is based on co-operation and mutual objectives. Therefore, a supply chain strategy that is going to use foreign companies on a lowest price only selection basis is unlikely to benefit from partnering. Therefore, the development of the supply chain strategy and decision to use partnering must be made simultaneously.

12.3.2 Partnering Development

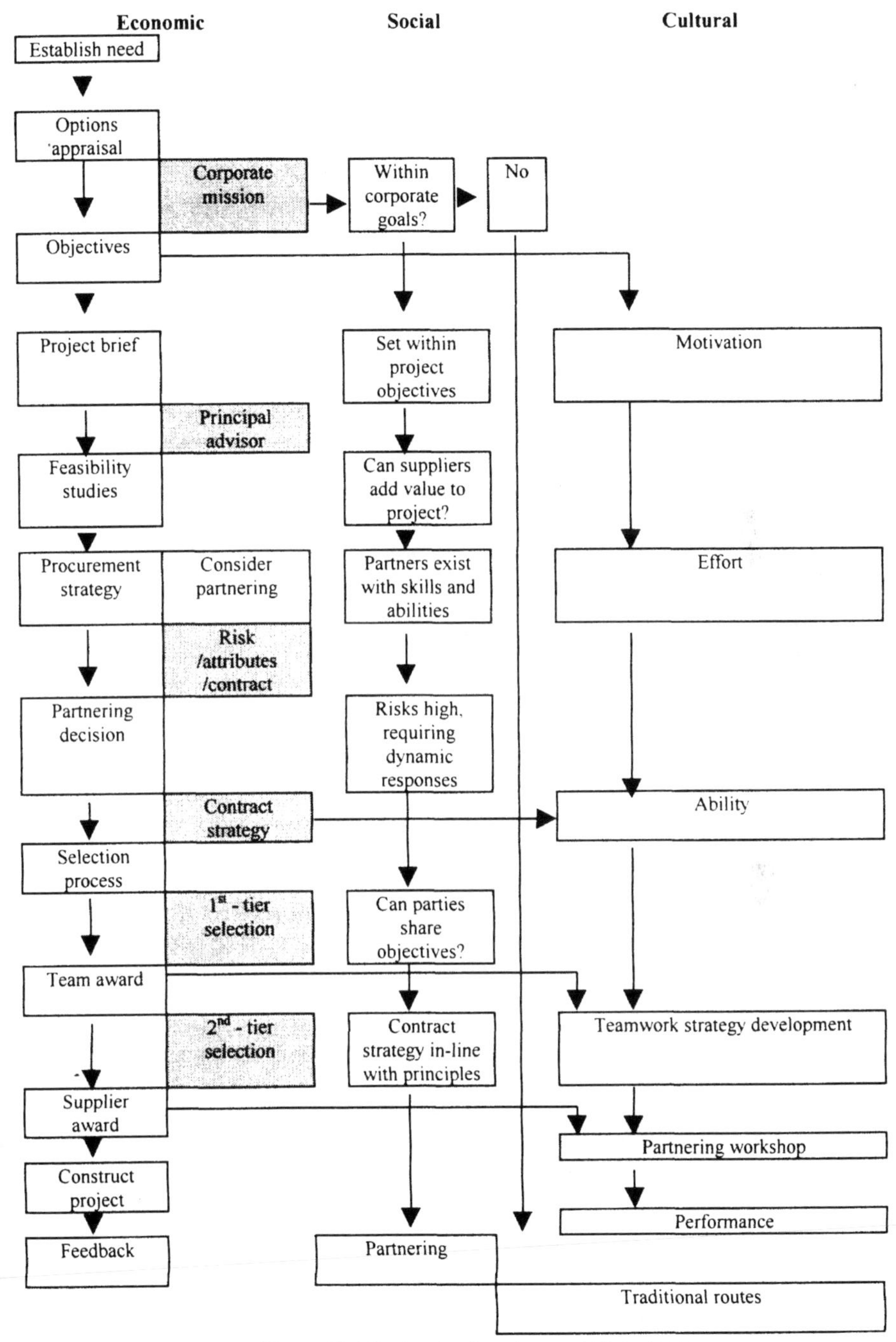

Figure 12.4 Partnering Development Flowchart

12.3.3 Partnering Considerations

Use the points below as a partnering checklist during the development of a procurement strategy.

1. Consideration must be given during the development of the project strategy to the following:
   - the decision to partner must be defined within the project objectives and overall development strategy
   - that partnering can allow for competition, but not adversarialism, on a number of fronts
   - whether partnering has the full support of all the project stakeholders
   - whether the benefits received on the definition of requirements, apportionment of risk, input to innovative products and designs and processes will enhance the project.
2. Consideration must be given during the development of the procurement strategy to:
   - the required level of competition
   - the use of pricing and payment mechanisms
   - the required ability to change by all organisations involved
   - the development of a risk management strategy.
3. If partnering is deemed beneficial to the project, then a strategy will have to be developed for the:
   - selection of individuals
   - selection of partnership organisations
   - benchmarking of results
   - appropriate managerial system to be installed
   - inclusion of individuals that fit within the system
   - clear defining of roles
   - senior level champions to co-ordinate organisation's involvement and provide coaching to own participants.
4. The project will have to have a clear definition of the technical, commercial and operational boundaries. This should be evident in the selection procedure for partners.

   Partnering organisations should:
   - be capable of learning from external sources
   - be able to learn from mistakes in an open and non threatening environment
   - have an in-house benchmarking and performance improvement policy
   - have a challenge-based paradigm
   - be willing to share objectives and personnel lessons.

   Partnering organisations should be selected by:
   - their comparability of culture and management structures
   - their comparability of business objectives
   - having a respect for people-based company culture
   - having an understanding of client's objectives, business and corporate strategy
   - their financial and corporate stability
   - their approach to risk management.

5. All projects will require a contract. The contract strategy should reflect the ideal of partnering and include:
   - the objectives of the procurement strategy
   - a reflection of the industries capabilities, skills, experiences and innovations
   - supporting the partnering charters
   - suitable pricing and payment mechanisms
   - an escalating dispute mechanism
   - a clear division of responsibilities
   - proper and fair apportionment of risk
   - a definition of project objectives and how their achievement will be defined.
6. During all stages of the process, efforts must be made to foster an environment that provides for relationships that:
   - are based on openness and co-operation
   - are capable of providing a positive working environment
   - are co-operative within working relationships to meet objectives and challenges
   - ensure all personnel understand partnering and its requirements
   - are based on continuous improvement in performance, quality and environment.

   Relationships should be underpinned by a managerial process that provides the following:
   - Identification: of clearly defined objectives that are embodied within the project strategy.
   - Stakeholder team: representing, where applicable, the various people who have a share in the required success of the project.
   - Monitoring: performance measurement should be used to ensure these are being met.
   - Charter: that embodies the objectives of the partnering principles set for the project. This should be embodied within the various parties' contracts.
   - Communication: not relying on the traditional communication hierarchies, clear principles should be set as to how communications are to be made between parties.
   - Value management: a system of ensuring design and commercial decisions are in line with the project objects and are maximising value for money.
   - Risk management: ensuring risks are identified and an appropriate management system is in place.
   - Continuous improvement: taking on the issues being raised in the management systems and monitoring results and instigating and changing management systems to ensuring improved results.
   - Dispute resolution: an escalation procedure to resolve problems.

Partnering Development Stages	
Development Steps	**Key Activities and Issues**
Owner Decision to Partner	▪ understanding partnering concepts and requirements ▪ suitable circumstances ▪ business needs and drivers ▪ evaluation and alternative strategies ▪ senior management alignment and commitment
Owner Preparatory Steps	Internal Alignment ▪ identify champions/project leaders ▪ business team/project team alignment ▪ owner competencies and role ▪ owner team Establish Alliance Contracting/Formation Strategy ▪ alliance design ▪ timing of selection ▪ contract structures ▪ remuneration terms ▪ selection process Alliance Contractor Selection Process ▪ establish selection criteria ▪ prepare selection criteria ▪ prepare selection evaluation plan
Alliance Partner Selection	▪ owner communication of intent to potential alliance contractor ▪ issue selection documents or initiate selection process ▪ evaluate responses and select
Alliance Development, Alignment and Commitment	Build Alliance Relationships ▪ apply facilitation, training, coaching and team building ▪ develop and apply communication processes ▪ apply/design other alignment mechanisms ▪ develop and initiate performance improvement innovation processes Jointly Develop ▪ project technical definition ▪ execution plans and programmes ▪ cost estimates ▪ risk analysis Finalise Works Contracts for Execution Phase Develop and Finalise Alliance Agreements ▪ project objectives ▪ principles of relationship ▪ project performance measures ▪ incentive scheme ▪ roles, responsibilities and decision-making ▪ dispute resolution Design and Establish Integrated Project Organisation Identify/Develop Common Processes and Procedures Build Relationships with Other Parties ▪ non-alliance companies ▪ external authorities

Owners Approval to Proceed with Project	
Develop and Sustain Alliance	Establish Team Delivery Targets Monitor and Modify Project Organisation Monitor Relationship Quality Continue ▪ performance improvement and innovation processes ▪ facilitation, training, coaching and team building ▪ building and sustaining relationships with others Monitor and Report Performance

Figure 12.5 Partnering Development Stages

12.4 KEY PERFORMANCE INDICATORS

In an effort to provide for a basis of comparing company and project performance, in the year 2000 the then Department of Transport, Environment and the Regions supported an initiative to produce key performance indicators (KPIs) for the construction industry. A set of 10 headline KPIs was produced reflecting the performance of main contractors in several construction categories.

The KPIs are a mixture of corporate performances and projects performance. They can be used within the procurement process to assess either the standing of a possible partner or set the performance targets for the project. As yet no similar performance measures exist for consultants.

This initial set was later expanded to incorporate ones specifically for the mechanical and electrical (M&E) contracting industry. The 10 headline KPIs encompass those aspects of service that were determined by clients of the building services industry to be the critical factors for the delivery of successful projects. These headline KPIs are:

1. Client Satisfaction – Design
2. Client Satisfaction - Installation
3. Client Satisfaction – Service
4. Client Satisfaction – O&M manual
5. Predictability – Cost
6. Predictability – Time
7. Defects
8. Safety
9. Productivity
10. Profitability

To measure a performance score, two major principles apply to each KPI chart.

1. In each chart the numbers closest to the bottom left-hand corner reflect the worst scores, and top right, the best performance.
2. Scores are read first by finding your score on the vertical axis ("Y axis) and following the grid line until it meets the "S" curve. At this point a vertical line is drawn towards the base line to assess your benchmarking position.

The following provides a step-by-step guide (Figure 12.6 relates):

a) Taking the example below of extent of defects on handover, supposing the client has scored the performance of 7 out of 10, the score is registered along the "Y" axis.
b) A horizontal line 7 is drawn until it reaches the "S" curve.
c) A vertical line is projected down to the x axis.
d) The bottom set of figures (0 to 100%) show the portfolio for the industry. The arrow meets at position 24 the benchmark score.

The benchmark score means that 24% of the building services contracting industry scored less than this and 76% scored 7 or above.

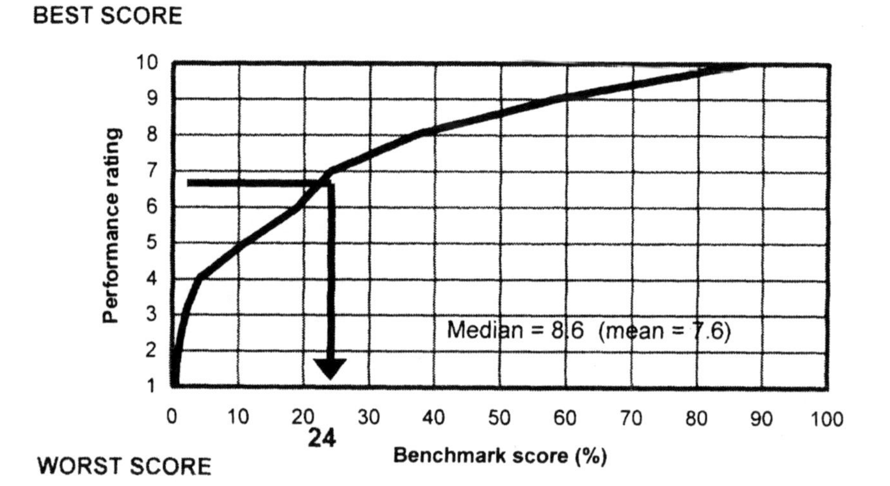

Figure 12.6 KPI Scoring Example

The median score on the graph is 8.6% and the mean score is 7.6. These are the industry scores for defects based upon research results. Each KPI graph was developed by independent research at BSRIA, who approached over 7000 clients of the industry for their perception of performance on projects in 2000. It is intended that the KPIs will be to produced and published on a yearly basis.

Each headline KPI is based on a specific question and a dependent methodology to work out the exact score.

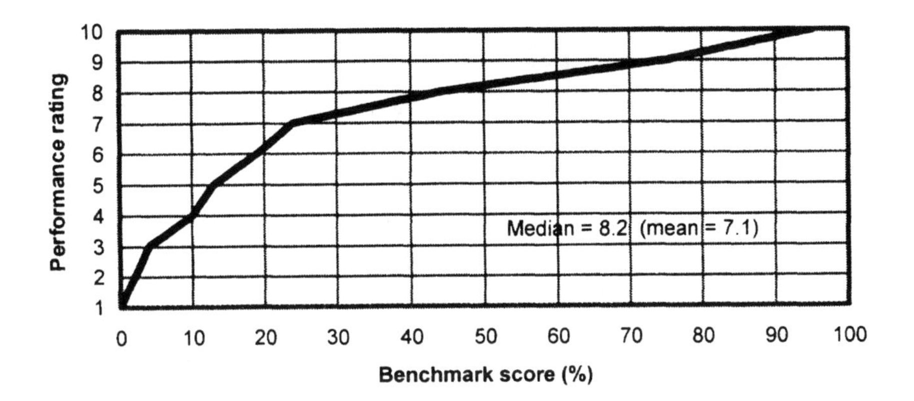

Figure 12.7 Headline KPI - Client Satisfaction – overall service

Question for Figure 12.7: How satisfied were you with the overall performance of the M&E contractor?

Worked example: if you score 7 out of 10 on your project (or 7 out of 10 across a number of projects) your benchmark score is 24. This means that 24% of the industry scored less than 7 and 76% scored 7 or more.

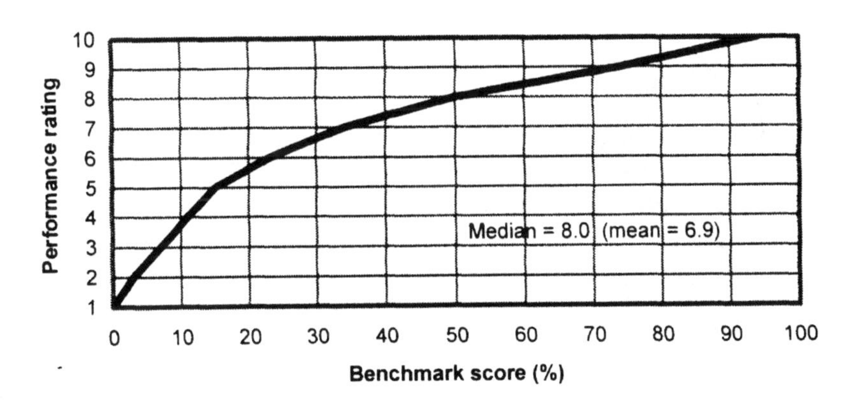

Figure 12.8 Headline KPI - Client Satisfaction with O&M Manual

Question for Figure 12.8: How satisfied were you with the quality of the O&M manual?

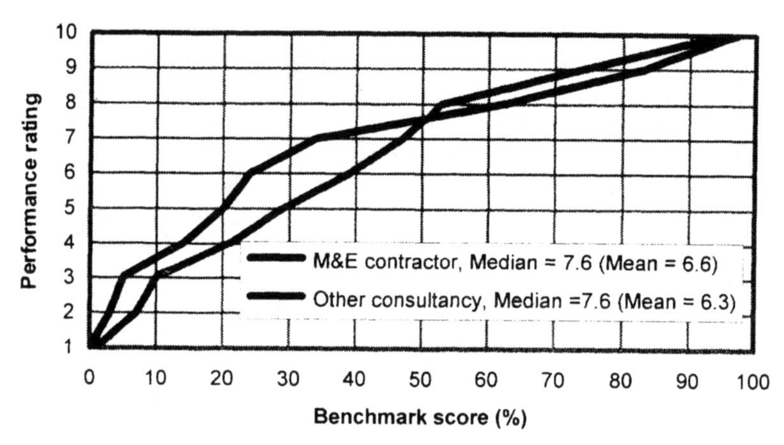

Figure 12.9 Headline KPI - Client Satisfaction with Overall Performance - Design

Question for Figure 12.9: How satisfied were you with the design services/design advice - overall performance, filtered by projects where the M&E contractor was involved in the design, versus projects where the M&E contractor did installation only (another consultancy undertook the design)?

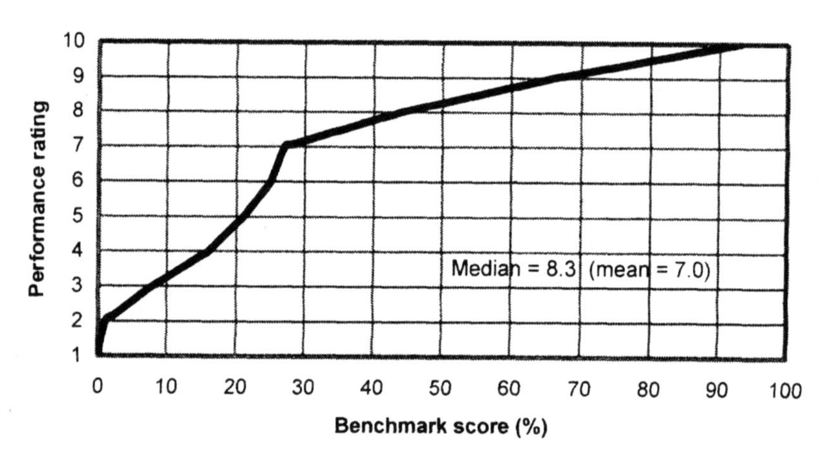

Figure 12.10 Headline KPI - Client Satisfaction - Installation

Question for Figure 12.10: How satisfied were you with the quality of the finished building services product/system, filtered by projects where the M&E contractor was only involved in the installation (i.e. eliminating the design element)?

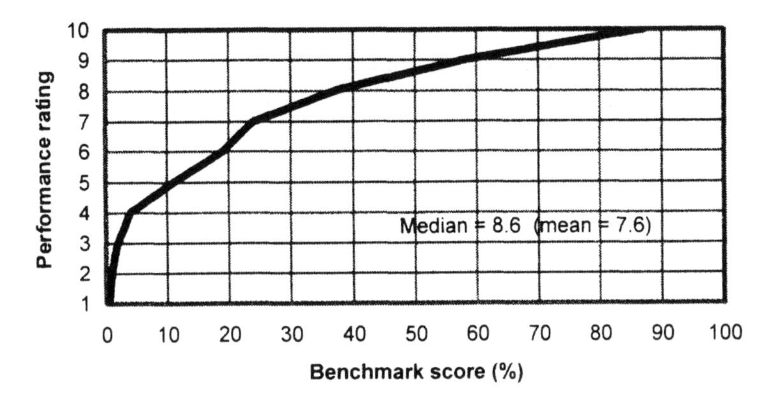

Figure 12.11 Headline KPI – Defects

Question for Figure 12:11: At the time of handover, what was the condition of the building services installation with respect to defects?

Key:

Score of 1-2	totally defective
Score of 3-4	major defects with major impact on client
Score of 5-7	some defects and some impact on client
Score of 8-9	a few defects and no significant impact on client
Score of 10	apparently defects-free.

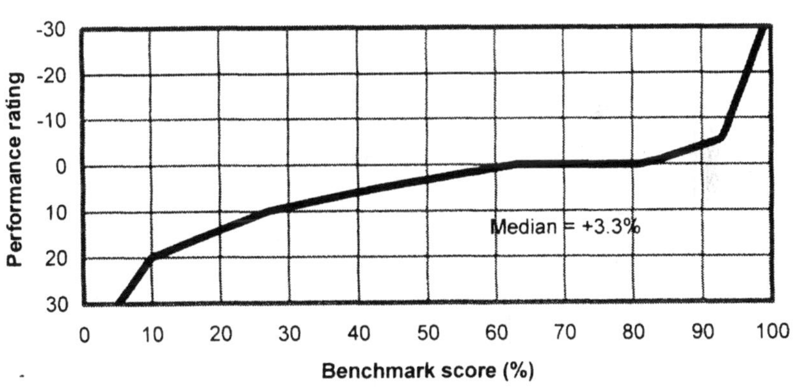

Figure 12.12 Headline KPI - Predictability of Cost

Question for Figure 12.12: What was the anticipated cost of the M&E installation?

<u>Worked example:</u>

Anticipated cost of installation was £500k

Actual cost of installation was £550k

550 minus 500, divided by 500, times 100 = +10% (i.e. the project was 10% over budget).

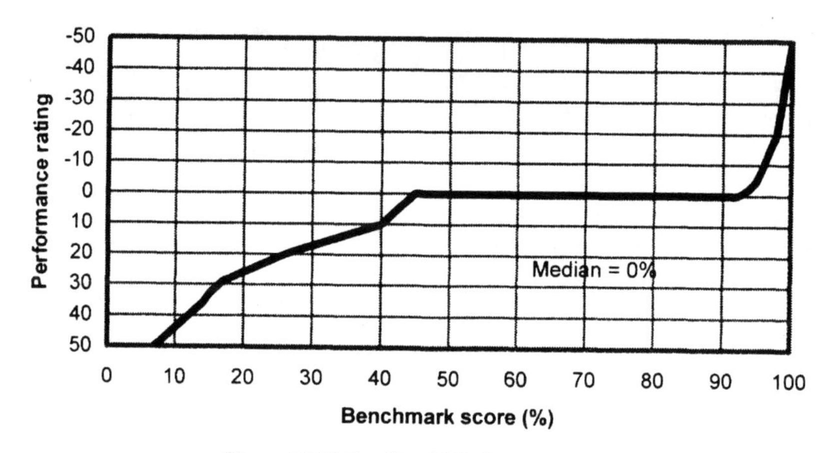

Figure 12.13 Headline KPI - Predictability of Time

Question for Figure 12.13: What was the total planned M&E installation time on site?

<u>Worked example:</u>
Anticipated time on site was 20 weeks
Actual time on site was 24 weeks

24 minus 20, divided by 20, times 100 = +20% (i.e. the project was over time by 20%).

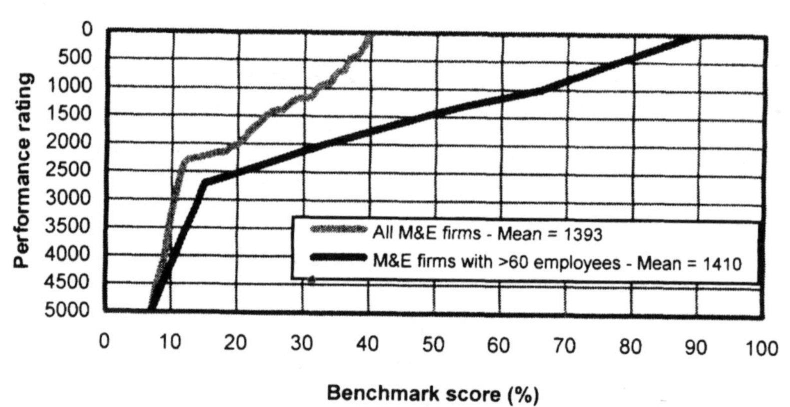

Figure 12.14 Headline KPI - Safety – Accident Incident Rate

Question for Figure 12.14: What were your reportable accidents in 1999? What is your company's staff level (all directly employed)?

<u>Worked example</u>: 2 reportable accidents, 36 M&E operatives, typical working week of 45 hours. As a rule of thumb, allow 47 working weeks per year.
36 (operatives) x 45 (hours worked per week) x 47 (weeks in year) = 76,140.

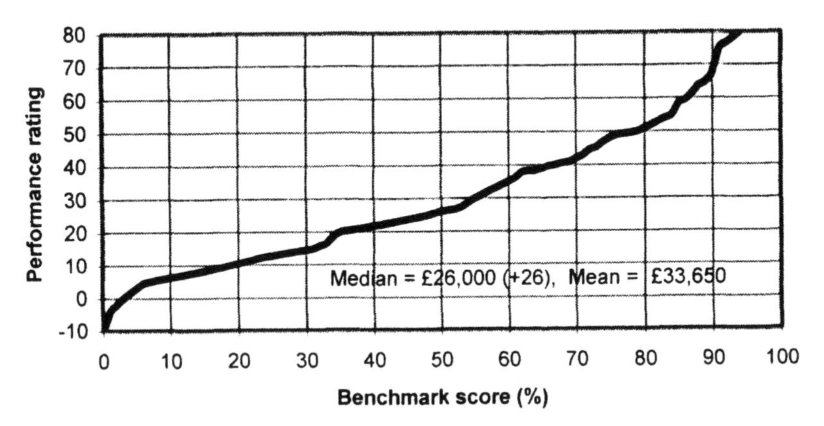

Figure 12.15 Headline KPI – Productivity - Value Added Turnover per M&E Operative

Question: What was your value added turnover in M&E contracting in 1999? If you also do contract maintenance/FM what percentage is this? What is your company's number of operatives (M&E installation only)?

<u>Worked example</u>
Value of turnover (excluding contract maintenance/FM) is £2,500k
Value of M&E subcontracted services is £1,200k
Value of goods supplied is £300k
Number of directly employed M&E operatives is 25.

1. Value added turnover is £2,500k - £1,200k - £300k = £1,000k
2. £1,000k divided by 25 (M&E operatives) is £40k, giving a benchmark score of 67%.

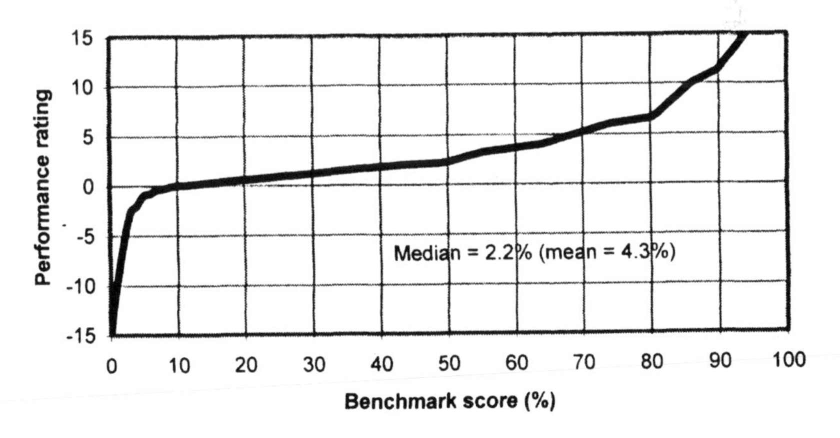

Figure 12.16 Headline KPI - Profitability

Question for Figure 12.16: What was your turnover in M&E contracting in 1999? If you also do contract maintenance/FM what percentage is this? What was your profit/loss from M&E installation in 1999 (i.e. excluding maintenance)?

Note: Value added turnover was not used in this KPI because figures would show an inflated profitability figure.

The data taken to calculate profitability also takes account of maintenance.

Worked example

1. Turnover from M&E installation (excluding contract maintenance/FM) is £2,160,000.
2. Profit from M&E installation (excluding contract maintenance/FM) is £42,000.
3. £42k divided by £2,160k times 100 = 1.9%
4. 1.9% gives a benchmark score of 41.

REFERENCES

1. BSRIA (2000) *Handbook and Guidance - Key Performance Indicators for M&E Contractors*, Bracknell, BSRIA Publications.
2. HM Treasury (1999) *Procurement Guidance No 3: Appointment of Consultants and Contractors*, HM Treasury Publications.
3. Scott, B. (2001) *Partnering In Europe - Incentive Based Alliancing for Projects,* London, Thomas Telford Publications Ltd.

FURTHER READING

For detailed information on the KPIs presented and to view the most up-to-date version, visit the website www.bsria.co.uk or e-mail kpi@bsria.co.uk

Chapter Thirteen

The Project Environment

13.1 PROJECT PROCUREMENT STRATEGY

The environment of any project is determined by its procurement arrangement. Traditionally, the appropriate arrangement has been determined by a simplistic approach to understanding the project objectives, with almost a complete lack of understanding as to how the chosen arrangement will affect the supply chain.

Most construction projects are large complex undertakings housing a myriad of cause and effect entities. Building services are one of the most important influencing entities. But as an element of the overall context of the project, its importance must be brought into line. Therefore, to devise an overall procurement strategy, the general nature of project procurement must be understood.

The chosen procurement arrangement will create an environment in which all specialist works must be undertaken. The boundaries of the environment, and the extent to which other parties, such as the designers, the client and the general nature of the industry can influence it, will be determined by the specific nuances of the main project procurement arrangement. To ensure that the manner in which the building services are to be installed is successful, an environment must be created that provides for the specific critical success factors. During the decision-making process of developing the procurement strategy, a balance must be made as to the environment needed for project success, versus that for the building services. It is doubtful that any particular arrangement would provide success for both. Therefore, the manner and extent of compromise must be decided upon.

The sections of this chapter outline the environment that is created by the primary contract procurement arrangement and that needed for the success of a building services installation.

13.1.1 Nature of Procurement Strategy and Arrangements

The overall end purpose of a procurement strategy is to select an arrangement that best suits the client needs in terms of meeting the cost, time and quality objectives. The problem with determining the best procurement arrangements for a client lies with determining what the client's requirements are and how these can be best served. Within all decisions, risk is one of the primary considerations. Each main project arrangement will mitigate and discharge risk in a differing manner. Taking into consideration that this text is written from the premise that current procurement decisions are inadequate in considering the ramifications of the building services supply chain, each arrangement must be reviewed in terms of risk and its ability to establish a proper supply-chain strategy.

There is no single approach establishing the optimum procurement arrangement, as each client and project will have differing objectives, project characteristics and influencing environments. As no two construction projects are alike, this raises the following procurement problems and the need to address how these can be resolved:

- describing or defining the procurement arrangement
- establishing the virtues of a procurement arrangement
- establishing the client's requirements
- matching the client's requirements to the procurement arrangement
- evaluating the performance of the chosen arrangement upon completion.

Successful procurement selection depends on matching the execution mechanisms with the client's needs, together with the characteristics of the procurement arrangement and the specific project attributes. This is difficult in construction for the following reasons:

- construction clients rarely identify precise requirements and constraints, due to time scales involved often changing business needs;
- the inadequacy of the procurement framework used for assessing the effectiveness of a chosen procurement path.

Identifying the client's specific requirements will be directly attributable to the experience of the client, while assessing the correct procurement framework is attributable to the ability and unbiased view of the project advisor. An experienced client may be able to form a more balanced view that an inexperienced one.

Procurement strategies can be divided into two main categories, traditional and non-traditional arrangements.

Traditional arrangements are characterised by the separate designers and contractors using a rigid design-measure-tender-build philosophy. Benefits to be gained from the use of traditional procurements systems are that they are well-known, documented and that the overall construction contractual arrangements are well-documented and tested. Although heavily criticised lately for producing poor results, it should not be understated that a tremendous number of successful projects are completed each year using this process. Approximately 60% of all construction projects are carried out using this form of procurement.

Non-traditional forms of procurement normally follow one of two patterns. They are either management-based, where the suppliers and construction teams are separate from the management of the project, or they combine the design and construction elements.

Lately, a second generation of non-traditional procurement arrangements have developed, where a lead organisation is responsible for all aspects of the project, from concept studies through to initial operation. These types of arrangements are only successful with considerable input from the client.

Although selecting a procurement arrangement is undertaken during a full strategic review of the project, a set of key factors are normally considered during initial evaluation. These can typically include:

- the type of client
- design input
- the management team
- availability of resources
- the type of risk evident
- the extent of risk the client can carry
- attitudes to risk taking
- responsibility
- legal requirements.

These major factors that are largely client specific must also be balanced with those that are project specific:

- the size of the project
- the continuity of the project
- previous problems
- the time available until completion
- the need for design change
- phasing requirements.

These reiterate the point that selection of the procurement arrangements must be specific to the project. Clients should adopt the most appropriate procurement arrangement to meet the needs of the project, but still be within their capabilities of involvement. Clients lacking the expertise to properly execute the direct involvement role are still able to appoint a surrogate client to act on their behalf, as in a procurement advisor.

Clients are at risk when selecting procurement arrangements, therefore all related issues must be reviewed and judged: client involvement is one of the most important covered. Non-traditional procurement arrangements normally require greater client involvement. The experience of the design team and the specific nature of the project will have a greater impact of risk onto the client than the selection of one arrangement or another. The most imperative aspect of selecting a procurement system is matching the client's needs and project requirements with the most appropriate arrangement. Inexperienced clients can overcome their shortfalls by the appointment of an appropriate and unbiased project advisor.

Each major set of procurement arrangements is now examined too understand how they affect and are affected by the building services element. It should be noted that the diagrams and relationships shown are not definite at all, each can be shaped and affected by the actual contract used. Instead, they show the general relationship between, rather than being specifically related to, a particular contract.

13.1.2 Traditional Primary Contract Arrangements

Traditional Procurement

The main characteristic of traditional procurement arrangements is the separati of design and construction, with the various designers and contractors carrying out roles which are defined and regimented by the various contracts and traditions enforced by the industry (Figure 13.1).

This category contains one main system - the conventional approach - with all other systems derived from its general principles, i.e. negotiation, two-stage, continuity and serial contracts.

The separation of design and construction causes traditional systems to be characterised by a sequential project delivery system. Thereby design must be largely completed prior to commencement of construction, hence its "traditional" tag. The designer designs and the contractor builds, with the responsibilities being strictly divided, even to the extent that the consultant is paid a fee plus expenses basis, with the primary contractor being paid on a competitive lump sum agreement. This itself causes problems with building services. The building services' team is immediately separated – both physically and contractually. A higher degree of co-ordination is needed, as is documentation to fully explain the actions required from each team member.

The system is highly criticised for the separation of the parties, with each working for a different master. While the designers are paid by a set fee, often determined devoid of any competition, the suppliers and contractors are pitted against each other in full competition. Supported by detailed prescriptive documentation, the contractors undertake only the work as it is described. Any deviation from this description results in claims for additional money and a long process of instruction issues.

As the price and manner in which the work is to be executed is determined only by the contractor, the designers and client have no opportunity to influence the supply chain or gain the benefit of any knowledge held by the specialist contractors. In sum, the primary contractor determines the building services' procurement strategy, other than the design.

Against these criticisms are the benefits of the system. Through the use of nomination, the client and designers are individually selecting each item and specialist contractors for the project. They are able to develop and execute the most elaborate of procurement strategies, each package being procured on its own merits and decision mechanisms. The ability to do so is limited by the designer's knowledge of the marketplace, the ability to develop such a strategy and willingness to accept the responsibility of the decisions. It must be acknowledged that a considerable amount of paperwork is required for each nomination package.

Furthermore, it still remains the responsibility of the primary contractor to ensure that all of the nominated packages work decisively together.

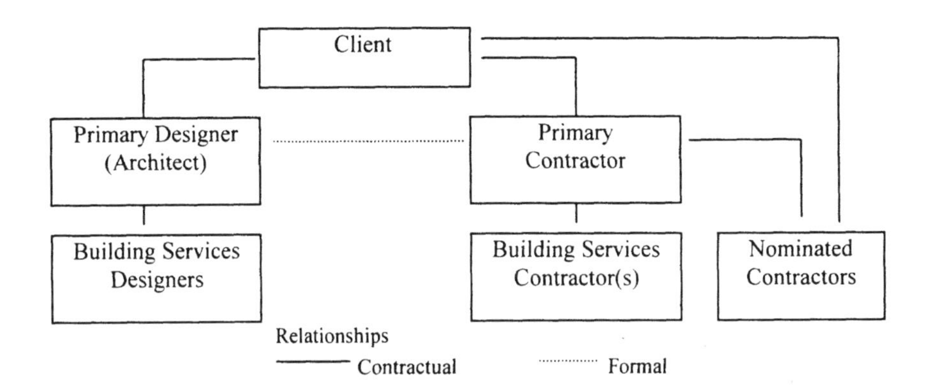

Critical Success Factors

cost:	lowest cost through competition,
time:	extended time due to sequential activities
quality:	requires considerable documentation to ensure quality standards are maintained
risk:	most transferred to contractor regardless of nature

Figure 13.1 Traditional Procurement Arrangement

The RIBA Plan of Works is the main dictator of the traditional procurement system. It sets out a sequential pattern of events characterised by four stages.

Preparation

- the inception stage, where the client establishes the need in terms of functionality, quality, cost and time, normally through the architect.

Design

- the appointment of a design team for the development of the brief, feasibility, outline design and scheme design;
- the lead designer, normally the architect, appoints the other consultants – this relies on the architect having in-depth knowledge of the building services requirements to appoint the appropriate party or parties;
- the system does not normally allow contact with contractors, but nomination and novation can be used for specialist works;
- design changes will incur cost and time penalties.

Preparing and obtaining tenders

- in order to operate successfully and enable the system to function, the design must be complete prior to the drafting of a bill of quantities;
- the primary contractor submits a lump sum fixed price for the works;
- all suppliers and contractors are contracted to the primary contractor;
- contractors must identify, accept and cost all risks;

- unless nomination or novation is used, the building services procurement strategy is determined by the main contractor only.

Construction

- the architect and designers play a viewing-only role, issuing instructions on behalf of the client;
- unless nomination is used, all issues of construction rest with the primary contractor and the various suppliers and specialist contractors;
- the building services team is separated, with designers working for one party and the suppliers and contractors working for another.

Providing the design has been fully carried out and the bill of quantities are accurate, the following benefits are obtainable from the conventional procurement method:

- a low capital cost driven by market forces, but is void of any whole life costing
- quantitative risk is eliminated
- competition is ensured
- post-contract changes are executed in a fair way and at reasonable cost
- clients are aware of the financial commitment.

A number of variations of the traditional arrangement exist. The most common is a two-stage tender. This is where three to six contractors are asked to bid on approximate documentation and submit details of expertise, resources and site organisation. A contractor is selected after an interview process and an analysis of their tender using a detailed price and quality assessment score. The chosen contractor then co-operates with the design team to give advice on design buildability. Advantages of this system include a reduction in preconstruction design times, together with preparation of tender documentation. As the price cannot be finalised until after the contractor is selected, the client pays a price premium. This must be balanced against savings in time and any advantages to be had from early involvement of the contractor.

Integrated Procurement Systems

All integrated procurement arrangements share the commonality of combining design and construction. Through this integration the benefits of accelerated time (due to the overlap of design and construction) single point responsibility and cost savings through efficiency are perceived to exist.

Although certain arrangements such as design and manage are considered to be integrated systems, they are not true systems as the management of design is carried out on a fee basis, with the works contractors carrying out construction work, therefore design and construction is not placed with any single party.

Often considered as a modern procurement arrangement, design and build is the oldest form of procurement in the UK, dating back to the historical contracting of master builders, who designed and executed the works using directly employed labour. It has enjoyed a varied resurgence in the past 25 years, by clients seeking a system that offered a fixed price for the project, together with discharging the majority of the project risk.

As the project team is an integration of both design and construction, a single communication channel exists between client and team, therefore contractual and functional relationships should be easier. But this simplicity is deceptive, as various constituent parts are complex and contain a number of pitfalls for inexperienced clients.

The problem stems from the setting of the brief, which is the only document the contractor will use to design and execute the project. Therefore the project deliverables and requirements must be highly detailed in a comprehensive and precise brief. Most briefs, however, share the normal pitfalls in their development - briefs are normally vague or badly written. Any client, other than the most experienced, will require an external advisor to assist in the brief development.

For building services in particular this is inherently difficult. Numerous decisions must be made regarding the building services design, which are dependent upon the specific design. While the generalities can be described, the detailed specifics often need to be worked out as the design develops. Therefore, design and build is a quandary for most projects, and is often unsuitable for projects with complex requirements. Moreover, beyond the setting of the brief, all aspects of the building services installation, from design through to equipment selection is carried out by the contractor. The only influence held by the client is through the performance requirements set within the brief.

The ability with which the contractor can deal with building services is determined by the nature of their own organisation. Generally design and build organisations are one of three categories:

1. **Pure design and build:** a complete and self-contained system where all design and construction expertise lies within one organisation, these firms are normally rare and usually operate in specific market sectors. Quite often they employ building services engineers directly and therefore have considerable expertise, often with specific types of installations.
2. **Integrated design and build:** normally a hard core of designers and co-ordinators, with additional and specialist expertise brought in, where more effort is needed by the contractor to fully co-ordinate the two divisions. It provides the advantage for building services in matching the designer's capabilities with the project requirements, but its disadvantage is the need for integration with other designers.
3. **Fragmented design and build:** design consultants are external and co-ordinated by internal project managers working for a main contractor. This arrangement normally manifests into ambiguity due to lack of expertise and the unfamiliar role of an architect working for a contractor. The same holds true for the building services' designers.

With regard to building services procurement, design and build offers an excellent opportunity. Notwithstanding the lack of ability for the client to influence the supply chain, the primary contractor is free to determine the procurement strategy for the building services. The contractor is able to instil a strategy that combines all aspects of best practice, from a fully integrated team to selecting the partners on a predetermined quality and price mechanism. The ability and willingness to break the work down into work packages is unlimited, only constrained by the contractor's own abilities.

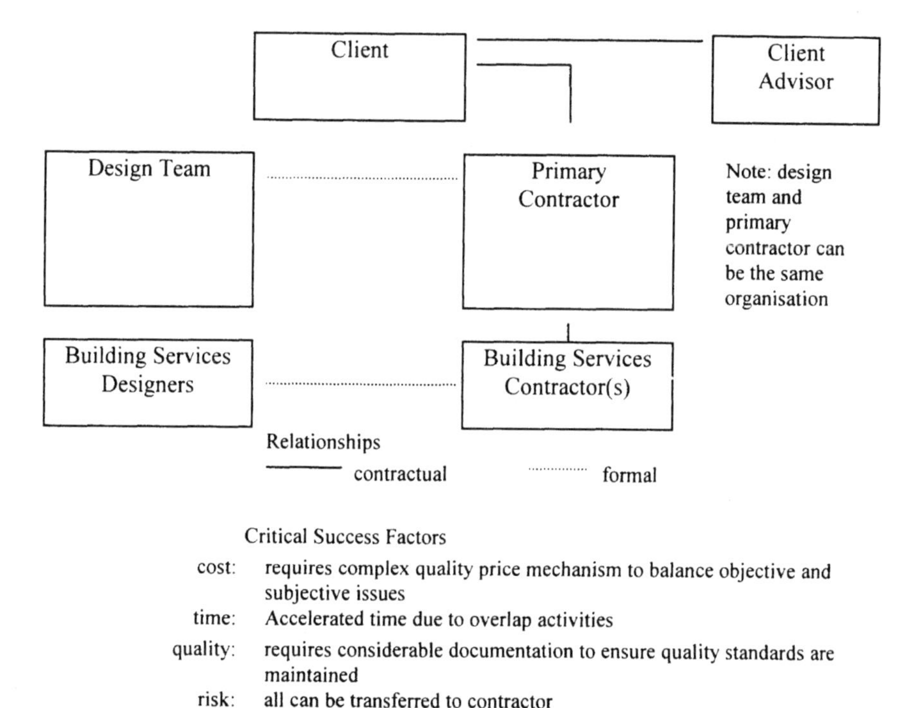

Figure 13.2 Design and Build Procurement Arrangement

There are a number of variants on design and build projects. These range from package deals, where the intention is for the client to procure a complete building. They are normally characterised by some form of systems building and therefore some compromises in client's requirement is inevitable. These then extend to more complex arrangements such as turnkey methods where the complete system is built and commissioned, often including the training of operation staff. These are most common for complicated industrial projects such as oil refineries or process plants. One variant that is more commonly used is develop and construct where the contractor is given a brief design from which they develop detailed drawings and specification.

There is no definitive system to implement an integrated procurement arrangement other than the three primary stages of determining the brief, determining the tendering arrangements and selecting the contractor. The sophistication of the project and level of competition required will dictate the precise system.

13.1.3 Management Orientated Procurement Systems

Two specific types dominate these procurement arrangements: management contracting and construction management. The main difference between them is the manner in which the suppliers, designers and contractors are engaged. Simplistically, under management contracting they work for the management contractor, while with construction management they are employed by the client, but supervised by the construction manager.

Management Contracting

The general procedures for management contracting are outlined in the standard contract document JCT Standard Form of Management Contracting. Having its own contract, management contracting is one of the oldest and most established of management-based procurement arrangements, although it is rarely used now in its original form (see Figure 13.3).

The arrangement has a number of unique characteristics, the predominant one being that the contractor is appointed on a professional basis, with their reimbursement being a lump sum-based fee, plus the prime cost of the construction works. Works contractors appointed and paid for by a contractor carrying out the actual construction. The designers are procured on a separate basis and are contracted directly with the client. The management contractor only provides the overall management of the project through the following services:

- preparation of the overall project programme
- material and component delivery schedules
- advice on buildability
- establishment of construction methods
- the detailed construction programme
- services and site facilities
- procurement advice on works packages
- a list of potential bidders
- preparing and organising tender documents
- placing orders with the works contractors.

The advantages to the client of using such a system are that they are able to influence the appointment of the building services supply chain without taking the risk in their appointment or management. The management contractor and the designer can be appointed at the same time, as can any of the works contractors. For building services procurement management contracting can offer these advantages, but is disadvantaged in that all risk of construction is passed to the works contractors, often through onerous contract conditions.

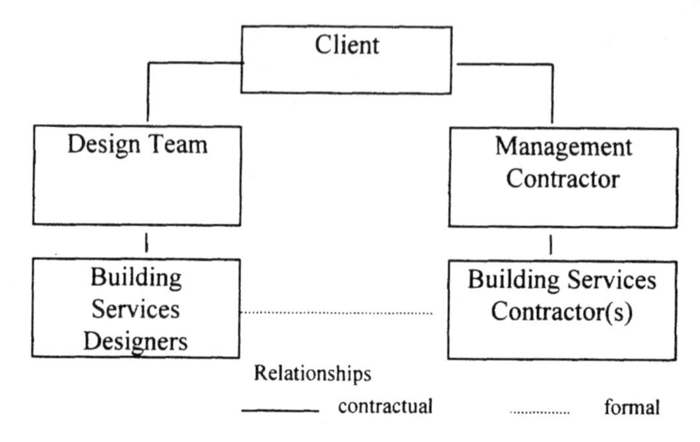

Critical Success Factors

cost:	medium, as cost is difficult to assess
time:	accelerated time due to overlap activities
quality:	variable
risk:	all can be transferred to contractor

Figure 13.3 Management Contracting Procurement Arrangement

Due to early involvement the contractor should be able to advise on design, buildability and quality together with cost to ensure the building meets the clients needs. To be successful requires the design team to enter into open dialogue with the contractor, therefore the client must ensure the team chosen are familiar and receptive to this form of management and involvement.

Other issues that affect building services' procurement associated with this arrangement include:

- a high degree of flexibility, allowing changes to be absorbed and rescheduled;
- contractors and designers can collaborate on new techniques as each can be appointed at the same time;
- the client must accept greater risk and responsibility as they are not protected by the management contractor who has been appointed on a professional basis, but it may advantage the building services contractor as they are able to deal directly with the client;
- management contractors must accept both management and construction risks;
- administrative effort is high ;
- work is divided into differing packages depending on elements and nature, allowing fine tuning of the procurement strategy;
- the contractor, client and design team work jointly to establish, procure and contract works contractors;

- the management contractor is precluded from carrying out works themselves to eliminate self-interest in commercial matters;
- the management contractor is paid on a certificate basis with each sub-contractor notified directly of the certified amount, thereby ensuring proper payment.

Construction Management

Construction management is very similar to management contracting with one very major difference. The construction manager, normally a contractor, is appointed as consultant and is on equal terms to the professional team, with the construction being undertaken by works contractors *directly employed* by the client. This is a very subtle difference, but its importance should not be glossed over.

Construction management is the only procurement arrangement where the building services contractors and suppliers can work directly for the client. It provides the most flexible and dynamic manner in which building services can be procured. The options exist to implement procurement from anywhere amid the spectrum of strategies; starting with a complete design and install package from a specialist contractor to a complete segregation of each system using distinct works packaging. The client is free to enter into a contract that may go beyond initial construction and may involve operating the system on a service basis or include a comprehensive maintenance programme. Furthermore, as the contract is between the client and specialist contractor, the payment mechanisms can be any combination of capital, whole life or operating cost.

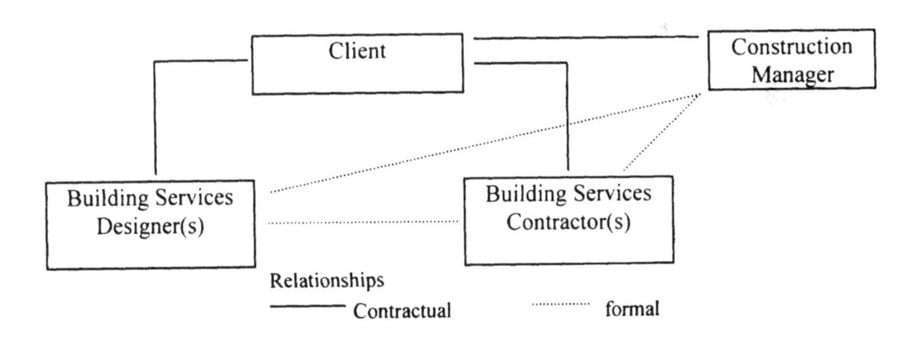

Critical Success Factors

cost: variable as management lump sum, but supply chain can be bid in a variety of formats
time: accelerated time due to overlap activities and ability to deal directly with supply chain
quality: assured due to ability to control each work package
risk: all can be transferred to contractor

Figure 13.4 Construction Management Procurement Arrangement

Similar to most management type appointments construction management offers similar advantages and disadvantages while raising similar issues. The construction manager's responsibilities are in line with their status as a professional manager:

- overall planning and project management of the works;
- assessment of buildability;
- cost advice, in conjunction with the quantity surveyor (if appointed);
- identification of statutory requirements;
- planning, management and execution of the construction phase, including works package management;
- informing the client on progress and required actions.

Generally product, time and cost are similar to most other forms of management contracting, including the typical criticisms.

13.1.4 Modern Procurement Arrangements

Along with the continuous development of the "traditional" procurement arrangement, a new generation of procurement strategies has emerged, driven by experienced clients seeking benefits from each member of the supply chain, while discharging maximum risk. Private industry and government bodies have sought improved methods of procurement, while reacting to the various reports that have highlighted the shortcomings of traditional procurement arrangements.

For the UK government, reaction to both the Latham report – *Constructing the Team* – and the Egan report – *Rethinking Construction*, have resulted in Prime Contracting for the Ministry of Defence and Procure 21 for the National Health Service. All government procurement is effected by the Best Value regime. But this is not a procurement strategy per se, but a benchmark of improvements within service delivery. In a similar vein private industry has developed similar strategies. Despite a plethora of titles – Procure 21, Prime Contracting, Project Jaguar and others – they all share the same basic core philosophies.

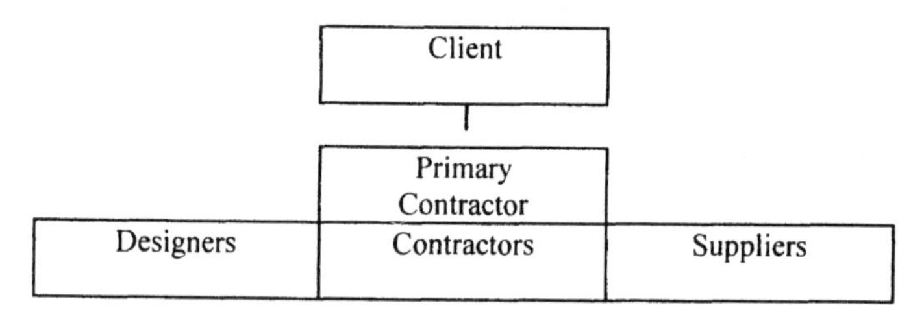

Critical Success Factors

cost:	single lump sum, competition is driven through innovation
time:	accelerated time due to overlap activities
quality:	should be high, as based on continuous improvement
risk:	all can be transferred to contractor

Figure 13.5 Modern Procurement Arrangement

The arrangement is based around the prime contractor, an organisation that heads a supply chain that have contracted together to deliver the project. The prime contractor could be any one of a diverse range of organisations, including contractors, management organisations or a special joint venture company. The prime contractor is selected from a short list held by the client and has pre-qualified based upon their management systems and, most importantly, the established supply chain links they have with industry. The main driver behind the arrangement is that the prime contractor would have a small well-defined supply chain that works on all projects together. The building services team is normally a key member of the project team, thereby heightening the level of teamwork and established lines of communication.

Working for a functional brief, the prime contractor is charged with designing and constructing all the works normally to a guaranteed lump sum price. In certain arrangements such as prime contracting, the contractor must also guarantee the operating cost of the building for its first three years of operation.

The main perceived advantage of these arrangements is that design and construction are integrated by a team that works regularly together. The team, using the best strategy available, procures the supply chain. It is perceived that by doing so the client is ensured of an innovative design solution, backed up with a guarantee of performance, both functionally and economically.

13.1.5 Private Finance Arrangements

Government-based private finance arrangements vary from other forms of procurement as the building or construction project may not be the focus of the contract. The partnership is normally set up to operate a building or similar entity and gains its money from the level of service provided, often back-to-back with the number of users. For building services procurement it does provide some unique procurement conditions. As the partnership is charged with running the building at a profit whole life cost and maintenance strategy become the critical success factors in procurement.

Private finance arrangements are characterised in the UK government through the use of a Public Private Partnership (PPP). This is a partnership between the public and private sector for the purpose of delivering a project or service traditionally provided by the public sector. Public Private Partnership is based on the premise that both the public sector and the private sector have certain advantages relative to the other in the performance of specific tasks. By allowing each sector to do what it does best, public services and infrastructure can be provided in the most economically efficient manner.

The above statements are a simplistic overview of PPPs. In reality they are complex business deals that often use a combination of private and public money, private and public buildings and similar distinctions over the people they contain. Most PPPs can be defined through eight characteristics. These are:

- Ownership
- Management control
- Financial/legal structure
- Source of investment funds
- Source and form of finance
- Safeguards for service quality (regulation)
- Employee representation
- Line of accountability.

Beyond the normal assumption that the joint venture company (normally referred to as a special vehicle company) set up to execute the contract would procure the building services installation in a traditional manner, PPP's provide the building services team with two potentially revolutionary means of procurement.

The first opportunity lies with the building services team forming part of the special vehicle. Under these circumstances the team would have to be a formal joint venture company developed by leading companies. The financial commitment for any project is extensive, thereby precluding any company other than a major organisation. In addition to the high financial returns, the services team would be responsible for the design, installation, operation and maintenance of the installation over its life. This involves higher financial risk and a joint agreement that all members accept any failings regardless of origin or responsibility. The reward for undertaking such a contract would be higher levels of profit together with a regular income stream.

The second opportunity for the building services team is by acting as a virtual company. The PPP joint venture company is able to procure the building services in any manner, but their expertise lies in financial economics rather than services technology. By forming themselves into a virtual company, the building services team can offer the PPP a comprehensive design and build procurement arrangement. By assembling the expertise around product types, the team can become specialists in particular requirements, such as hospitals or schools. The same advantages exist for the team as the first scenario without the risk or need for investment capital.

13.2 GENERAL APPROACHES TO BUILDING SERVICES

In the previous sections of this chapter the various procurement arrangements that are in use today have been outlined in the context of how they affect building services procurement. At this stage of the text it is appropriate to summarise the main issues that must be considered during procurement and prior to considering an improved methodology.

13.2.1 Problems with Current Procurement Arrangements

Current procurement arrangements, other than construction management, require the primary contractor to procure the building services on the contractor's terms of reference rather than those of the client. A lack of available expertise in most contractors, together with the short time periods associated with tendering often cause a disparity between what is needed for successful procurement and what is available.

The principal issue was stated in Chapter 1:

The traditional belief normally held within procurement strategy is that the abilities of the main contractors to procure the correct specialist contractors is due to their strategic abilities of dealing with the supply chain. It is normally assumed that the main contractor holds a dynamic relationship with the market, with the consultants determining strategic objectives from the client. However that assumes the subcontractor has no strategic influence and will therefore make the best possible choice purely in terms of resources and methods. Furthermore, the critical success factors for project success under this arrangement are compatible with the subindustry delivery process.

In reality, however, main contractors are management contractors subletting almost all the construction works to domestic subcontractors on documentation that does not recognise this to be the case. The main contractor in fact has little strategic control over risk or price on the projects, as their role has become that of a co-ordinator only.

Setting aside the previous discussion in this chapter, from the view point of building services, the differences between alternative procurements frameworks are down to:

- the arrangements for determining cost
- the process of selecting the contractor
- the role of specialists
- the process structure
- the conditions of contract.

The role of the specialist can be used to differentiate between procurement arrangements – construction management and lump sum arrangements are equal, other than the specialist being able to participate in the design phase with construction management.

13.2.2 The Key Issues

Five main arrangements exist for the procurement of building services:

1. **De facto systems:** this type of arrangement is not strategic, but a result of the main contract procurement strategy. They could be crudely referred to as pot luck systems as the design and construction are separate with the

services procurement strategy being determined by the main contractor. Although this is the traditional manner in which services are procured, the increasing levels of technology no longer make this an acceptable methodology.

2. **Separated systems:** using the principles of supply chain management each technical system could be procured separately. Design and construction can also be split down further. However, this requires a high degree of sophistication by the client, compatibility with the main construction strategy and a high degree of co-ordination and integration skills by the contractors and consultants.
3. **Integrated systems:** the responsibility of design and construction rest with one organisation. These systems are typically used for simple installations where the systems performance can be precisely defined. A strategy must be developed that allows early selection of the contractor based on their abilities in design and construction.
4. **Management-oriented systems:** the overall management of design and construction is given to one organisation, who co-ordinates the others. Typical arrangements in general construction include construction management and management contracting, with the contractual arrangement being the main difference between them. However, both these strategies can be applied to building services.
5. **Product-based systems**: the project is delivered as a product, where the client uses the facility on a pay-per-use basis. The supplying contractor is charged with designing, building and financing the entire project, normally including its end maintenance and management. The responsibility of ensuring the project meets the needs of the client remains with the contractor. Private Finance, Public/Private Partnerships, design-build-operate, and many other connotations can be used. These strategies are likely to increase in the future with a greater emphasis on corporate PFI and serviced facilities.

The difference between these strategies results from the way risk is apportioned between the parties and the amount transferred from the client, including:

- **Financial risk**: product-based strategies allow complete risk transfer to the contractor, including future maintenance and operating costs. The client only pays a user charge, often linked to the operating function of the building.
- **Buildability**: the importance of the constructor's knowledge base is now acknowledged. This is combined with an assessment of the likely complexity of the project. Each individual element must be considered, as simple structures can often have complex services requirement, e.g. call centres.
- **Timing**: the actual time to develop, construct and commission the building will depend on the sequence of individual trades. Timing of traditional arrangements cannot be modified without considerable penalties. Management-based arrangements allow greater flexibility, due to the direct relationship between contractors and client. The overall

length of time available for the project and critical completion dates will affect the end choice of arrangement.

- **Organisation**: the client's ability to, and desire for, control of each party within the construction team differs between arrangements, as does the accountability and responsibility. The client's experience and need for direct involvement needs to be considered.
- **Commissioning:** the disparate tasks of balancing a designer's model, client wishes and contractors comprehension of these, lies with the commissioning engineer. Ideally this party should always be independent, but this ideal is only compatible with certain strategies.
- **Co-ordination:** the ability to properly integrate the various managerial and technical subsystems must be held by a company that can balance the various conflicting requirements and act as arbitrator in clashes of requirements or ideals. The degree to which clients must become involved differs with each strategy. The importance of becoming involved should not be misunderstood or underestimated.
- **Contract:** the final element within the procurement strategy is the choice of contract. Tailored contracts allow the final proportioning of risk and clear definition of responsibilities. Gone are the days of the adversarial tabloid used to transfer all risk to the lowest level. Modern contracts are based on proper apportioning and fairness. However, procurers can use a quirk in law to their own risk purposes. Contractors must use due care and attention when designing and installing an installation. However, a designer needs only to exercise reasonable care and skill: an interesting thought when developing a procurement strategy.

The influence building services has on the cost, quality and programme of a construction project is a well-known and often cited fact. What is less well-known, is the relationship between the procurement strategy of the entire project and the procurement strategy of the building services. Furthermore, the strategy developed for building services must consider a number of variables that are independent of the main project yet will directly affect project performance.

Unlike the main contract no definitive relation types can be drawn up. Each factor can be arranged and executed in any manner thereby creating an endless assortment of possibilities. The final decision should not be viewed as any particular relationship, but as a business plan in which decisions must be made as to how best to make use of the resources available.

Each system needing to be procured can be viewed as a make or buy strategy. Is it preferable to buy the service – from a combined design and build specialist – or "make it", by purchasing each individual component and labour force to assemble it? Further consideration can be given to the labour element: employ directly or employ through an agency (as a subcontractor). This entire process can be undertaken for each individual resource until a definitive strategy is developed. The final chapter in the text provides guidance for undertaking such an analysis.

REFERENCES

1. CIBSE (2000) *Guide to Ownership, Operation and Maintenance of Building Services*, London, CIBSE Publications.
2. Egan, J. (1998) *Rethinking Construction,* London, DETR Publications.
3. Latham, M. (1984) *Constructing The Team*, London, HMSO.
4. Masterman, J.W.E. (1992) *An Introduction to Building Procurement Systems,* London, E&FN Spon.
5. Turner, A. (1990) *Building Procurement*, Basingstoke, The Macmillan Press Ltd.
6. Wild, J. (1997) *Site Management of Building Services Contractors*, London, E&FN Spon.

Chapter Fourteen

Design Responsibilities and Programmes

14.1 ALLOCATING DESIGN RESPONSIBILITY

14.1.1 Industry Practice

Design is one of the most fundamentally important aspects of any installation. Building services are a unique element, within which design of both its technology and methodology of installation is undertaken. Every element of installation, including the plant, is a unique arrangement, specific only to that building and particular design. Although a "designer" is responsible for the design, its actual execution transcends many levels. From the specialist engineering contractor to the manufacturer, each is involved in some element of design. To be successful, each of these elements must be fully co-ordinated and achieve its desired performance.

Sufficient mechanisms must be developed on a project to ensure that each member of the building services team understands the following:

- who is designing each element
- what is the extent of their design
- who is to receive the design information
- who is co-ordinating which element of the design
- what is the design programme.

Failure to address these questions adequately in establishing the management and procurement strategy in previous projects has often resulted in disputes, delays and quality problems.

To achieve the full level of co-ordination need, it is preferable that a procurement arrangement that favours teamwork is established. Although the methodologies that are outlined further within the text will assist, high levels of communication are needed. Traditional attitudes to construction, normally occurring when traditional procurement arrangements are used, tend to develop the relationship into a contractual environment, where only formal communications and definitive boundaries of responsibility become endemic. This problem is further compounded by the lack of a clearly identified and independent party responsible for the co-ordination. Within the traditional procurement arrangement this would be either the architect or the main contractor. But neither party has traditionally held the appropriate expertise nor has there been an industry-recognised standard for co-ordinating the works.

Successful co-ordination requires the mutual agreement of all parties involved with design, from the principal designer through to the component manufacturer.

14.1.2 Allocating Responsibility

Allocating clearly defined boundaries around each party's area of responsibility sounds ideal, but the reality is far from it. Increasing complexities of technology involve greater levels of design, often created by specialists whose specific undertakings will not be understood by other parties. Furthermore, the increasing use of packaged plant and protection over copyrights limits the specific details of certain areas of design.

Within the context of procurement, it is critical to establish the level and area of responsibility at the strategy stage. But as the design is yet to be developed, particulars cannot be given. Instead, a management framework must be developed that can evolve through the project.

There are three key stages involved with correctly allocating design responsibilities:

1. **Work package allocation:** the functional performance must be stated, either as detailed specifics or as general descriptions within the design brief. This is a critical stage, as clear identification of the functionality will affect all other decisions. Although this stage is normally undertaken within the main brief, design is an intuitive process, resulting in additional items or specialist items that may require additional briefing. Further, the brief may only state generalities, requiring the establishment of certain functional parameters before the appropriate allocation of responsibility can take place.
2. **Detailed responsibility allocation:** once the overall design strategy for the project is complete, a work breakdown structure can be completed. At this stage the specifics over plant requirements, systems design and key contractors will have been identified. Detailed design responsibility can now be allocated to the most appropriate party, in line with the management structure and procurement strategy.
3. **Responsibility statement:** the allocated responsibility must be clearly stated. Although the use of a detailed allocation matrix is ideal, some form of mechanism must be used to instruct the responsible party. At the early stages, the principal designers and contractors would have this allocated in their tender or similar documentation. It is during design development and procurement of the second-tier specialists that documentation becomes greater and more varied. Use can be made of drawings, specification clauses, contract clauses or design scope statements. Each must be

interlinked and compatible, with drawings especially needing a precise definition of what information is to be included.

When developing a procurement strategy it would suffice to use a simple allocation matrix to initially define the design logic as shown in Figure 14.1:

Procurement Stage Design Responsibility Matrix	**Work Package**					
	Comfort Cooling	**Heating**	**Power Circuits**	**Solar Shading**	**Lighting**	**Controls**
Responsibility						
Client	FP	FP	FP			FP
Architect				FP/CD	FP	
Lead Designer	CD	CD	CD		CD/DD	
Lead Mechanical Contractor	DD	DD				
Lead Electrical Contractor			DD			
Specialist Contractor		DD radiant floor		DD		CD/DD
Plant Supplier	DD-ahu's					
FP = Functional Parameters CD = Concept Design DD = Detailed Design	ahu = air handling units					

Figure 14.1 Work Package Design Responsibility

For the vast majority of projects, the above matrix would be self-evident. It should also be noted that the chart merely shows the main party responsible for a particular function. In the true nature of design all parties would interact and be involved, i.e. as a typical example the lead services designer would be involved in all aspects of functional determination, but may not be primarily responsible for it.

During the decision-making process the matrix becomes invaluable for several reasons:

- the criticality of each specialist becomes evident – predetermining when and how they should be procured;
- the immediate identification of each party's responsibilities is possible;
- the interrelationship between the works package and design is evident ;
- the necessary lines of communication are identifiable;
- with the addition of a dates column the initial design programme can be developed;
- the matrix can then be adapted for the detailed assessment of design responsibilities.

Allocating design responsibilities goes beyond the mere apportioning of drawings and calculations. Several activities can be allocated to different parties or shared. Each activity will need to be considered during the formulation of the procurement strategy as each will be affected by the procurement arrangement and workload level accepted by the party. Beyond the identification of the key party responsible for the activity, there are a number of sub-activities that need to be allocated during the detailed design stage that will affect design responsibilities, including the activities shown in Figure 14.2 below.

Activity	**Detailed Considerations**
Plant selection and procurement:	Allocation of plant selection and procurement should be based on: ▪ capital plant – over a designated value ▪ critical function plant – providing a critical function requiring particular considerations, e.g. UPS ▪ high-maintenance plant – where detailed whole life and maintenance regimes must be considered ▪ packaged systems
Selection of specialist contractors and designers:	When specialists are appointed, either independently or through another party, decisions will need to be made concerning: ▪ the timing of the appointment ▪ the responsibility for information provision ▪ the extent of information provided to and from them ▪ communication procedures ▪ acceptance of the specialists design
Role of commissioning engineer:	Ideally the commissioning engineer will be an independent appointment whose activities will include: ▪ reviewing the design responsibility strategy ▪ commenting on the design programme ▪ reviewing design ▪ advising on commissioning requirements
Hand-over information:	Clear allocation of handover information responsibility includes: ▪ the person responsible for overall co-ordination ▪ the required format and layout of information ▪ agreement on the scope of information

Figure 14.2 Design Activities Allocation

14.2 INTEGRATING DESIGN AND CONSTRUCTION

Calls for the integration of design and construction have been directly linked to the problems in the industry with regard to the discontinuous nature of the procurement process failing to provide an integrated product of design and manufacture. This same discontinuous nature of work and requirements for flexibility has also been encountered by other industries. By the adoption of integrated design and flexible manufacturing systems, these industries have achieved what the construction industry is seeking - high productivity with added value, while at the same time offering quality and good design.

The key to understanding such a subject must take place at the manufacturing level - in the case of construction at the specialist engineering contractors level. The actual manufacturing of the building has now passed to the subcontractor or specialist. As the subcontractor is now the hirer of plant, the paymaster of labour and the buyer of materials, the knowledge base previously held by the main contractor has disintegrated due to its concentration on management rather than manufacture.

One of the key concepts in transforming construction into a manufacturing industry is the transference of design responsibility. This role, normally carried out by the architect, has been historically criticised due to the quasi-relationship between the legal requirements of the contract and the lateral thought process of determining the client's requirements. The segregation of knowledge bases within the industry has meant the architect has been unable to fully consider all aspects of the design. This has led to poor buildability and a high degree of custom components.

In order to gain maximum benefit from the manufacturing process and to facilitate production, the designer must either be fully conversant with the manufacturing process or have continuous access to a knowledge base that can deliver the required input. Previous research into integration has concluded that the key issues affecting integration are:

- simplified design
- standardisation
- repetition of design elements
- effective communication of design information
- dimensional co-ordination
- compatibility of elements and material tolerances
- design input.

It is fairly clear that most of these are based at the manufacturing level, but are a direct result of the detail and/or concept design.

Standard components should lead to greater economies within building costs and can allow more efficient production, both in manufacture and site assembly. Standardised detailing has been directly attributed to the overall efficient speed of the American construction industry. Flexible manufacturing techniques utilising a common base design can be developed. Designers would then be able to produce concept designs in a similar base fashion, but with each standard component tailored for specific reasons to create an overall unique design.

The use of standard components and design solutions appear to be the greatest advantages of integrated design and manufacture. Customers of the building will also be able to make readily identifiable decisions on product quality and performance in a structured and systematic manner, with all the implications of their decision known to them. The whole construction process can then become a linear manufacturing process of clear decisions based upon simple designs. The result will be a high value product of simple design.

One of the problems within design has been understanding the complexity of the resulting design, which should be considered at the concept stage. Research by other industries has been able to measure complexity and has led to great improvements in design, most notably in the electronics industry. By understanding the complexity at any stage of the design - from the vaguest of concepts to the finality of detail - the designer should be able to understand the cost and buildability implications. By reducing the complexity the component should become more reliable, at a lower cost and therefore of a higher value. The resulting simplicity of the component will incur fewer steps in manufacture and design and fewer discontinuities in the system. The value will remain intact and have less chance of deteriorating or being discontinued.

Another problem with design will be that of single point fixation - the point where a designer's mind can become fixated on a specific solution. By segregating the concept and detail stages the designer is not able to have lateral input from specialists, also, the specialist will fixate only on the solutions they can offer, e.g. an electrical specialist will be unable to offer advice on electrical systems. At the point where a specialist is appointed, the design will become fixed to his particular area of expertise. Obviously this could be solved in one of two manners: either the design process must be carried out using a democratic process in which various design solutions must be considered for comment or judgement, or the concepts architect must continue as design co-ordinator and must balance the intents of each party. However, this contradicts the current belief that the architect's knowledge base is not great enough to make balanced decisions.

In order for the industry to adopt integrated manufacturing philosophy, it must begin to re-evaluate business methods and concentrate on the notion of work being a product or specific service to be sold and marketed. This requires a social change in the methodologies and ideologies of how business should be conducted. The future for successful subcontracting lies in subcontractors finding niche markets or specialist design and install packages. By forming the services or products into branded market specific items, customers will select by value rather than price.

However, the success of this type of strategy has been hampered due to varying reasons, most often cited as:

- limited finance
- lack of marketing / business skills

- limited time development
- staid marketplace.

Concentration on manufacturing and standardisation of components should allow site performance levels to increase. A large amount of a building's value is deteriorated by low site efficiency levels. BSRIA's own research cites that site performance efficiency is around 30% and inefficiency in production processes reduce production out-turn by 25%. The success of the United States and Japanese industries is attributable to their repetitive working, using simpler and fewer operations.

By standardising the processes, site operations will become simpler and less labour-intensive. However, the current site operative will need to be retrained to learn transferable assembly skills. This will require a digression from current craft training. American construction is largely based upon semi-skilled labour using high levels of mechanisation to site-assemble prefabricated components. This should result in higher levels of production, together with higher financial returns for the operatives.

The separation of manufacturing and design has plagued other industries, but has been resolved by the use of innovative manufacturing and management techniques. All of these have the same common attributes of flexibility, design and manufacturing integration, together with balancing standardisation with custom detailing. Research into alternative manufacturing techniques has shown that the use of such systems in manufacturing have not stifled design creativity or direction by myopic manufacturing views - and are not just suitable for detailed design. The following are easily identifiable techniques, all of which are in common use within other industries:

- simultaneous engineering
- rapid prototyping
- team design
- just-in-time
- batch processing
- design for manufacture (DFM)
- design for production (DFP)
- design for piece-part assembly (DFP)

Procurement practices currently being exercised must change to allow early involvement of the construction and manufacturing industries. However, specific research into the appropriateness of this provides a complicated picture. Current procurement principles of partnering and value engineering rely on this early involvement to maximise value and improve the buildability. By doing so, the design co-ordination of individual components and overall design concepts can be improved. The end result is an elegant solution to the traditional muddle of contractual responsibilities. Traditional forms of procurement do not allow this integration of the individual design teams due to varying objectives and approaches.

The procurement arrangement options need to be amended to take into consideration that for overall quality, the specialist will dictate on value, performance and the cost of the building. In order to maximise value, the specialist must be allowed the flexibility and scope to input their knowledge at the design stage, to allow manufacturing to become efficient and economic. Most main contractors will subcontract works into convenient packages to minimise risk to themselves and optimise risk transfer to the specialist, rather than concentrating on value optimisation. The main contractor can maximise his profit by distancing the price and value of the individual services. At the same time the minimal price for the work is paid to the specialist, due to this distancing of the main contractor, thereby allowing no additional revenue for investment into modernisation or innovation.

14.3 PROJECT PROGRAMMES

14.3.1 Programming Design

The current trend in construction procurement is to find an arrangement that delivers superior levels of time, cost and value for a specific construction project. In doing so, clients have begun to favour arrangements that overlap design and construction activities, on the assumption that overlap savings will deliver cost and time benefits. This assumption is based upon the notion that concurrent design and construction can be used. While this is a logical assumption, the programming of design for building services is often affected by the overall design programme set by the architectural design.

However, design programming and design management are currently of poor quality. Furthermore, the ability to programme design, and therefore establish the correct overlap with a construction programme is not understood. Current guidance documents, such as the two NEDO documents produced in the 1980s, only give rudimentary times, without providing detailed buildup of the established time, and therefore any possible design/construct overlap cannot be established. By establishing the required overlap, the most appropriate procurement arrangement can be determined.

Establishing design decision points will also allow the establishment of the proper period required by the design team to implement the supply chain. Knowledge bases within the supply chain need to be captured and introduced at the appropriate time for efficient and quality decision-making. Therefore, the most important element in any design programme may not be the overall duration of the activity, but rather where it occurs in the sequence, together with the specific nature of the activity.

Procurement

Savings in time can be achieved by overlapping the design and construction phases, and non-traditional procurement routes allow this to be more easily carried out. Most procurement guides and textbooks reiterate this point, showing graphical representations of how design and construction can be overlapped. However, with no reference guide to design programmes - especially for building services - nor any manner of determining them, the validity of these claims must be called into question. This paralleling also has an affect on the design methodology and activities, often to the detriment of the project.

Therefore, the available overlap between construction and design will limit the number of successful procurement arrangements available. Subprogramme issues affect overall programme times within the procurement stage. For example, building services installation programmes are often limited by the extended deliver times for large plant.

The strength of modelling design systems lies in their ability to apply to all levels as each element of a building is a subsystem requiring its own design and procurement process as a self-contained item. Good management, and therefore good programming, must identify these elements and the information sources, normally from the specialists, and co-ordinate their timely input. The timing of this input by a specialist is critical. Designers and suppliers of specialist services should be brought in at an early stage in the design process. Their expertise in advising on the best solution would ensure that their own contribution could be fully integrated into the project design and construction programme at the appropriate time. By not appointing the specialist in time, usually caused by the chosen procurement route, a hiatus in design can occur. Thus the design programme and programme route are inseparable and must be considered together.

Current Programming Techniques

Generally, five methods of design programming dominate current thought, namely:

- network charts - critical path types
- bar charts
- procurement schedules
- information-required schedules
- information transfer schedules.

Only the first two can be fully considered as programming methods. The remaining three rely on input from the contractor to state the required information and its required date. The result is programming chasing rather than planning, with design often being forced into an out-of-sequence logic or resulting in an unrealistic amount of information produced in short time scales.

Information transfer schedules can also be criticised as the designer normally develops them from either best guess or subjective reasoning. Problems with this type of programming stem from it being concentrated on the production of the drawing only and treating each drawing as an independent task. The information analysis and synthesis stages of design are ignored.

Procurement schedules, while a valuable component, should not be relied upon for design purposes. Procurement schedules are based on the construction programme together with expected delivery times. They ignore the cyclical nature or logical process of good design. They can assist in developing a design programme by showing the latest completion dates, but should not be a substitute for good programming.

The effect of other elements on the building services' design must also be recognised. Most buildings consist of groups of common elements known as technology clusters, based upon the technology and performance criteria of the various elements. Typically a building can be divided into seven such clusters, namely:

1. sub-structures
2. structure
3. envelope
4. service core/risers/plant
5. entrance and vertical circulation spaces
6. horizontal surface finishes
7. vertical surface finishes.

By organising the project into such clusters, the correct information from specialists and consultants can be input at the correct time and level, but demonstrate that building services is a distinct element requiring a specific programme.

The importance of these clusters also lies with the call for multi-disciplinary design teams. As the design appears to be in segregated stages the feasibility of using this type of management style is possible. The time periods are short, which allows intense working, while solutions can be developed in isolation. What is of interest to designers is that the concept phase can remain untouched and developed in isolation.

Clusters allow design to become systematic, with all required inputs becoming integrated into a final solution. If this observation is true, that design can be split into lumps based upon a set criterion, then design programming can be based upon this. The individual tasks would not require programming, other than stating the required information inputs per cluster. Instead, programming would centre around the co-ordination of key stages and the definition of the required clusters.

Architectural Design Fees

The architectural programme dictates all other programmes. To understand its implications the general methodology used by architects to plan their work needs to be understood. The RIBA's *Architect's Appointment (1983)* breaks the architects work into specific stages (see Table 14.1).

If the fee stages are proportional to the man-hours required to complete the phase the framework for a design programme exists. It is interesting to note that the stages break down into fairly equal units suggesting that each phase of design takes a relatively equal amount of effort and therefore time. Detailed design of the building services cannot take place until work stage D is substantially completed. It should also be noted that if production information is not carried out until stage F and G, theoretically any overlap between construction and design cannot commence until 75% of the design phase has been completed.

Work Stage	**Fee Proportion %**	**Running Total**	**Stage Deliverable**
C	15	15	Outline proposals - sketch plans
D	20	35	Scheme design - sketch plans
E	20	55	Detail design - working drawings
F G	20	75	Production information - working drawings
H J K L	25	100	Site supervision

Table 14.1 RIBA Work Stage

Research carried out by Nicholson and Naamani (1992) into commercial buildings showed there was little relationship between fee income and the cost of providing the required services for a specific project. Their research showed that projects between values of £1-3 million showed a steady level of staff costs of £30,000 (1989 prices). It can be reasonably assumed that man-hours also remain constant. If this is so, then it is assumed that design programmes remain constant for a given type of building, regardless of its size.

Design Management

Part of the reluctance for non-use of design programming is due to the nature of the architect's contract with the client. There is no requirement in the standard RIBA *Architect's Appointment (1983)* for the architect to programme their work, other than to 'prepare an outline timescale'. Section F of the *Architect's Job Book* states that architects should develop a register of drawings and schedules required for the project. The interesting point about this is that it is recommended to occur at Stage F of design, that is, after detail is completed.

Preparation of design must be directed towards facilitating on-site progress, as construction is directly affected by design coherence and timely communication of information. The compatibility of both design and construction programmes needs to be established from the outset of construction. Numerous research projects have found there was a direct correlation between the speed of a project and the flow of well co-ordinated design information. Therefore, the components of a design programme must not only establish the overall feasibility of an overlap, but also ensure that the method of individual pieces of information, such as those required for building services, are co-ordinated in a timely manner.

This aspect of information delivery is currently becoming more complex due to the involvement of specialist contractors. Therefore, the design programme does not merely have to deal with the consultant's information outputs, but must also be able to co-ordinate and state the requirements of the specialist's contributions. Continuity of design is severely affected by the timing of appointments: research shows site operations were severally affected by delayed documentation due to the inconsistent sequencing of design and specialist input/information.

Provisional Design Programme

Based on previous research by this author, a provisional design programme has been developed based upon the analysis of previous projects. Although very rudimentary it does provide a framework for understanding the nature of architectural design and its implications for a procurement strategy.

The result of the research has identified nine key stages of design development that were common to all projects.

1. **Concept proposals:** initial concepts based upon pure design principles, typically the most creative part of the design. Often these were concentrated on the plan layout since, with industrial buildings, lorry-turning and goods transfer determines the success of the development. The specific duration of this sequence typically only took between 3-6 % of the total design man-hours, but the specific time period varied considerably. The affect of planning regulations and the clarity of the brief were influencing factors.
2. **2D concept check:** the site was specifically laid out with the initial concepts cross checked to ensure that the logic of the building design and site layout were compatible.
3. **3D concept development:** this stage develops the concept into 3D sketch drawings where both elevations and floor plans are cross-assembled to determine the overall logic and feasibility of the design.
4. **2D horizontal co-ordination:** with the specific building design developed in the previous stage, the site is specifically laid out and the feasibility of the building design is checked against site constraints. Although some general discussions over possible services design may have already take place, it is only when the building envelope has been configured and agreed that engineering design can begin.
5. **2D dimensional co-ordination:** this stage commences the working drawing stage with the layout of properly dimensioned and drawn plans of floors, elevations and cross-sections. These are laid out to scale, taking into consideration actual construction arrangements and dimensions.
6. **3D dimensional co-ordination:** at this point the design programme becomes complex due to the multi-sequencing of differing design programmes. Sufficient information is produced at this point to allow structural, civil and certain major elements to become designed in parallel. Towards the later stage of this phase three other sub-categories of drawing become highly influential. These are drawings that, by the nature of the project specifically fulfil one of three functions:

- project specific drawings
- programme dependent drawings
- procurement specific drawings.

At the completion of this stage sufficient engineering design and the necessary drawings for both plant and specialist engineering works have been completed. The detailed procurement of the project can now begin.

7. **Finishes layout:** this stage, together with stage 6, form the largest area where time is spent within the programme. They are also the two stages where the largest quantity of drawings are produced. The finishes layout includes different forms of drawing from detailed joinery drawings, and schedules of finishes, through to conceptual drawings for contractor-designed elements.
8. **Component layouts:** most programmes showed that these types of drawings were produced at the later stages of the design programme, but were not specifically sequential. Their detail and the number produced was directly related to the specific requirements of the design.
9. **Construction supervision:** although not specifically design orientated, the last stage is the on-site supervision of the executed design. It is also one of the areas in which the largest part of an architect's time is spent.

Design Programme								
Stages								
Initial Concept			Secondary Concept			Drawing Production		
1	2	3	4	5	6	7	8	9
concept propose	2D concept check	3D concept develop	2D horiz. coordin.	2D dim. coordin.	3D dim coordin.	finishes layout	comp layout	super
Programme Time								
4-6%	2-3%	4-7%	2-4%	10-20%	15-25%	15-25%	5-10%	17-30%
project commence			engineering concept commences		engineering detailed design commences; procurement overlap			

Figure 14.3 Design Programme

The sequential ability of the design development is upheld by comparison of the actual times taken within the project programme studies. It showed that parallel working by more than one individual only occurred during the later stages of the design programme where sequencing of design is possible. This occurs after stage five where grouped sets of unrelated drawings are produced.

This design programme begins the process of developing a framework for its further development as a useable programming technique. Furthermore, it allows the establishment of correct fees for a given project as the fees stages have been clearly identified. The developed programme has made considerable progress in identifying key design stages and the percentages of time taken for each stage.

14.3.2 Programming Procurement

Developing a procurement programme is the combination of developing a suitable strategy, while considering the actual implications of any decisions. Programmes consist of two main types and a third related programme, which follow the various hierarchies set out within the work breakdown structures:

Strategic programme: each procurement arrangement, whether for the main project or the building services, will dictate a general logic on the basis of which the principal designers and contractors are selected. This logic will set the key dates, for both the earliest and latest times when the principals need to join the project.

Tactical programme: once the principals are on board and design has sufficiently progressed, the second-tier suppliers and contractors need to be procured in line with the overall strategy.

Purchasing and resources programme: although materials purchasing and resources programmes are not normally considered in the strategic nature of procurement, key suppliers and items need to be identified early on. Major plant items or specialist equipment may have critical programme implications, warranting their early order or direct ordering by a principal party.

The main contract procurement arrangement and the specific selection mechanism used will limit the above programmes. Competitive tendering requires detailed design documents and therefore an extended time period to prepare the tendering documentation. The design programme will thus dictate the procurement programme. Time can be expedited through the issue of negotiated or partnering arrangements that require only sufficient information to appreciate the project. The legal significance of the document should be diminished through this approach. The decision over the appropriate mechanism will be based upon the client's objectives of time, cost and quality.

Programming is about time. Based on stated objectives time can always be compressed but would normally carry a cost penalty. To achieve a programme in a given time other mechanisms could be used, such as prefabrication or modularisation. As the main contract programme sets out the overall timeframe, the building services programme is often an issue of reverse engineering.

Other key issues that should be addressed by the procurement programme, as they may affect the overall procurement strategy to be adopted, include:

- **critical dates:** these are dictated by the main contract programme, usually for partial handover, testing or sectional completion;

- **plant deliveries:** since major plant items are often on extended delivery times, these need to be checked and programmed-in, based upon the construction programme;
- **building access dates:** these are either for sectional handing-over during a refurbishment, or are critical access windows for heavy lifting/plant movements;
- **heavy lift schedule:** all major plant requires heavy lifting and moving. Access for doing so is often limited by the building envelope programme and efficiency of crane hire;
- **assessment times:** bids, specification and drawings must all be assessed correctly, which requires time;
- **tendering times:** these vary, but the most notable documentation recommends three weeks for simple supply, four weeks for when work is premeasured and 6 weeks for tenders requiring design work. Tendering is discussed in detail in Chapter 15.

14.3.3 Programming Construction

Construction programmes for building services have for a long time suffered from the poor planning abilities of the main contractor. Even now, when known to be the critical path item on nearly all projects, building services is still limited to a single line on the master programme.

Construction programmes are based upon the logic that is required to construct the building. The overall programme is normally set by the clients key end date. The main contract procurement route further affects it. In this competitive arrangement, the main contractor may have formulated the programme without reference to the building services team, or shortened it to gain a commercial advantage.

To satisfy a constricted programme, and assuming the building services are on the critical path, requires a change of strategy. This can be enabled through procurement on two different levels: either through a change of strategy, combining the design with construction to increase the design/construction overlap (but be careful of the design programme required!), or by changing to a technical-based strategy that makes use of off-site prefabrication and modularisation. Another alternative is to adjust the work breakdown structure and select the critical elements on a non-competitive basis, thereby limiting the design and tender documentation required.

The procurement strategy must respect the needs of construction. But similarly, the construction programme cannot call for a delivery or installation schedule that would be either impossible to fulfil (due to long lead-in times of manufactured items), or absurdly high labour levels.

The strategy should not be developed until the construction programme has been developed and times for critical plant items have been ascertained.

REFERENCES

1. BSRIA (1997) *Allocation of Design Responsibilities for Building Engineering Services TN21/97*, Bracknell, BSRIA Publications.
2. Hawkins, G. (1997) *Improving M&E Site Productivity TN 14/97*, Bracknell, BSRIA Publications.
3. Marsh, C.J. (1998) *Design Programming - the procurement paradox,* unpublished MSc Dissertation, Nottingham, Nottingham Trent University.
4. NEDO (1983) *Faster Building for Industry,* London, National Economic Development Office.
5. NEDO (1988) *Faster Building for Commerce,* London, National Economic Development Office.
6. Nicholson, M.P. and Naamani, Z. (1992) *Managing Architectural Design- a recent survey*, London, Construction Management and Economics, Volume 10.
7. RIBA (1983) *Architect's Appointment*, London, RIBA Publications.
8. RIBA (1983) *Architect's Job Book*, London, RIBA Publications.

Chapter Fifteen

Tendering

15.1 TENDERING PROCEDURES

Tendering is the tactical process of selecting a party to provide a service, product or item of work. More recently it has been incorrectly associated with price-only competition. Regardless of the mechanism for selection, some type of offer and acceptance process must take place. The particular methodologies are often dictated by factors external to the project. The final procedure adopted and decisions made within the process must be in line with the procurement strategy and arrangement adopted.

It is a tactical process as the strategic decision over possible partners and the scope of work has already been made. It could also be viewed as the business or commercial process of procurement. To understand tendering within its proper context, any type of tender should be viewed as either a formal or informal procedure.

Informal procedures are not precisely defined and therefore it is difficult to give precise guidance for or recommendations of best practice. Informal tendering is both a traditional process and modern one. Traditionally, consultants and even contractors were hired based upon local knowledge and the social circles that existed within local business. In particular this is the manner in which most consultants obtained work. This type of selection has had a renaissance lately through partnering, where parties are selected based on recommendations and "soft-based" criteria such as quality and reputation.

Formal tendering procedures are usually characterised by systematic procedures normally set out in guidance documents that accompany contracts. The various procedures, set by the Joint Contracts Tribunal (JCT) dominate the industry. Furthermore, external influences can also dictate the manner in which the offer and acceptance is carried out. Government work must respect the local standing order, central government procedures and the formalities set by the European Commission with the EEC Directives.

15.1.1 Procedures

Procedures vary greatly and are often dictated by either: tradition, government procedures or accountability requirement. The specific procedure adopted will depend upon:

- the type of tendering being undertaken
- the number of tenderers selected
- contract conditions
- whether it is a single or two-stage tender
- the accountability required
- the expected value of tender
- the location of tenderers (local or international)
- the location of the project
- the procurement arrangement
- the selection mechanism (quality/price mechanism or cost only).

Generally, all tendering procedures follow a similar set of main activities, with the minor activities being dictated by the above factors, as outlined in Figure 15.1.

Major Activities	Minor Activities
Tender methodology	▪ Responding to factors ▪ Developing strategy ▪ Programme
Tender documentation	▪ Programme for production ▪ Documentation to be included
Selecting tenderers	▪ Pre-qualification invitation to possible tenderers ▪ Assess responses ▪ Interviews ▪ Independent checks
Tendering procedures	▪ Co-ordinate documentation ▪ Issue tenders ▪ Issue amendments
Evaluating tenders	▪ Site and external checks ▪ Assessing documentation ▪ Inputting to quality/price mechanism ▪ Interview ▪ Award

Figure 15.1 Tendering Activities

Tender Types

Each tender will be based upon a selection methodology that is in accordance with one of the four general methods. Each of these reflects different levels of competition. There must be compatibility between the tender type chosen and the weighted assessment mechanism used to compare the tenders. All four types of tenders stated can be used for either the services of contractors or consultants, or the provision of materials.

Open tender: Public bodies procuring bulk materials are normally the only users of open tenders. The cost of sending out unlimited tender documents normally excludes it from the detailed tender typically used for service provision or complicated supply times. It is, however, very useful as a first stage process to gain maximum interest from possible suppliers.

Serial tender: Serial tenders are normally used for a series of projects or for a project with uncertain quantities, such as a maintenance contract. The tenderer submits rates against approximate quantities or for a typical project. These are then used as the basis for a series of contracts. Tenders often state +/- percentages against the stated rates for the final number of contracts or quantities completed.

Competitive tender: The procurement of most items in construction, and particularly building services, is dominated by competitive tendering. A group of companies, typically up to six, prepare a fixed price, lump sum cost for completing a set of works. The submitted bids are normally only assessed on a price basis as the quality of the work, defined through the documentation and quality of the contractor, should have been assessed prior to being invited to tender. Although considerably maligned in recent years, competitive tendering can provide effective results as long as the following is observed:

- tenderers must be comparable in size, location and ability
- sufficient time is provided to complete the tender
- work can be properly described within written documentation
- the tenderers' responsibilities are clearly defined
- adequate time is available for the preparation of tender documentation and a considered response from the tenderers.

Negotiated tender: Negotiated tenders can either be single-stage or two-stage affairs. With single stage, a particular organisation is chosen and through negotiations agrees the various particulars for a project. The singe contractor is pre-selected either through past experiences with another party or through some form of prequalification procedure. A two-stage tender invites a limited number of organisations to submit details for a project, often including basic rates, percentage mark-ups and management information. Using the predetermined weighted assessment criteria, a tender is then invited to a second stage where the price is either negotiated or based on the original rates against a bill of quantities or schedule of works.

15.1.2 Timing of Appointments

There are two critical aspects to the timing of tenders. Firstly, tenders must be sent out in accordance with the procurement programme, allowing sufficient time to agree all details and allowing sufficient lead-in time for the parties once the work has been awarded. Secondly, the actual tendering period must be sufficient to allow the tendering parties enough time to fully understand and put forward a suitable tender offer. Inadequate allowance for either period will create unnecessary risks for both parties.

The time allowed for the preparation of a tender should be commensurate with the complexity and type of plant and site, the nature of work to be carried out and the type of contract involved. No definitive guidelines are available for correct tendering times, but the Code of Procedure for Single Stage Selective Tendering issued by the Joint Contracts Tribunal (JCT) states the following guidance:

- for projects involving defined and measured work – four weeks
- for projects involving elements of design – six weeks.

Although the above times are for complete construction projects, given the complexity of building services these should be used only as a guideline. Other guides, such as those produced by CIBSE, state that up to three weeks is adequate for simple projects, with additional time being given for larger or more complex projects.

Within the tender programme sufficient time must be allowed to answer queries raised by tenderers and the distribution of answers to all parties, regardless of the source of query.

The above times are for the actual preparation of the tender response. Prior to this sufficient time must be allocated within the precontract programme for the selection of appropriate tenderers. Appropriate parties should have been assessed initially during the preparation of the procurement strategy. The availability of suitable parties will have dictated the choice and methodology for tendering. Competitive tendering cannot be used if only one party can provide the necessary plant or service!

15.1.3 Documentation

Tendering documentation must completely describe the works involved with the project. For building services this must include the architectural elements and sufficient structural details to understand the building envelope's interrelationship with the services. Insufficient description of the project, or inadequate details of the required work to be undertaken, is the greatest source of disputes. The quality of the service provided is dictated by the quality of the tender documents.

The purpose of tender documents is to allow the tenderer the opportunity to put forward an offer to carry out the works described. The offer needs to be based on the objectives of the tender documentation, which are to provide or state:

- the basis of an understanding of the work required
- a commercial basis for valuing the works
- the contractual conditions to be used
- payment methodology and procedures
- design roles and responsibilities
- commissioning responsibilities
- programmes for procurement, main contract and building services
- the scope of other building services work packages
- the maintenance programme and strategy
- the person responsible for co-ordination and their scope of works.

To meet these objectives the tender documentation should, as a minimum, consist of:

- detailed scope of works
- instructions to tenderers explaining tender return dates and information required
- programmes for both the building services and main contract
- the minimum period of validity of tenders
- the contract to be used – if the contract is standard then the optional clauses and particulars need to be stated, or the complete contract is to be included
- architectural and structural details
- design responsibility – in the form of a matrix showing the responsibilities of all works packages involved with building services
- plant schedules
- whole life cost requirements
- minimum maintenance requirements
- spares and commissioning attendances to be provided
- management personnel to be provided
- operations and maintenance manuals requirements - numbers of and format
- design approval procedures
- a health and safety plan and precontract risk assessment
- a complete description of the works including drawings, specifications, calculations and design statement.

For all projects it is advisable to hold a tenderers' meeting during the tender period. The purpose of the meeting, held simultaneously with all tenderers, is to allow the clarification of any queries and the expectations of the client. If the project is already designed it provides the designer an opportunity to explain the reasoning behind the design. If the building structure is already under construction, or is a refurbishment the meeting should be combined with a site visit. By having the project explained in detail and all queries answered simultaneously, the tenderers gain a greater appreciation for the project and are more likely to submit complete tenders.

Although a site visit does not form part of the tendering documentation it is the single most important act that must be undertaken to ensure that the works described in the documentation are set within their proper context. If this is not undertaken than the tenderer is unable to fully appreciate the project. For consultants a visit to the client's business fulfils the same function, especially where some industrial process must be designed for.

One compulsory piece of documentation that must be included and returned by the tenderers is the pricing document. This document is either a detailed price schedule, or, less commonly, a bill of quantities. The increase in complexity of building services has not been met by quantity surveyors thereby resulting in a decrease in the use of bills of quantities.

The pricing schedule breaks the submitted tender amount into a detailed breakdown for each work section. The schedule should describe each element of work explicitly, ensuring that the tenderer will make due allowance within the stated price.

Beyond the statement of capital costs for the stated works, the pricing schedule could, depending upon the nature of the work package and procurement strategy, include provision for:

- a statement of anticipated whole life costs
- a maintenance cost schedule
- commissioning and capital spares costs (if required)
- pricing options for alternative systems, plant or service provision
- a schedule of overheads and profits
- a schedule of daywork rates, including plant and materials.

Note that if whole life costs form a major part of the project, then a schedule of assumptions needs to be provided, together with a statement as to how interest rates, time periods and maintenance strategy are to be determined. It is best to consider including a predetermined model with the tender.

The priced schedule must total the tender price submitted. The additional items do not need to be carried forward to the total, but instead act as a menu of price options.

15.1.4 Tender Submission

Historically, contractors have been forced to submit fully compliant tenders with no scope to submit alternatives. This is wrong.

A tender submission should be seen as the opportunity for the tenderer to put forward their best possible proposal to execute the work. A suitable compromise is to ask for compliant tenders with alternatives submitted as possible options. This permits innovation yet ensures that comparable tenders are received. Although submission should not vary widely from the stated requirements, sufficient freedom

should be allowed. Asking for only compliant tenders foreshortens the opportunity to learn of the contractor's experience or alternative approaches.
The submission should be a clearly defined objective statement as to how the tenderer is to approach the work. It should provide for sufficient information to assess the commercial, managerial and technical approaches to the project. It should not contain detailed information about the company. Sufficient research regarding the company should have been undertaken at the time of deciding the tender list, in order to ensure that the list is composed of suitable tenderers.

As a minimum the tender submission should contain the following:
General project approach statement, comprising:
- management structure
- labour histogram and staffing methods
- design programme and methodology
- cash flow or milestone schedule
- general method statements.

Agreement to contractual documentation, comprising:
- bonds
- warranties
- general contract conditions.

Commercial proposal, comprising:
- pricing schedule
- programme, showing both design and construction
- schedule of equipment suppliers and subcontractors.

Alternatives proposal, comprising:
- full details of alternatives, including quality certificates, outline design or method statement.

15.1.5 Evaluating Tenders

The submission of the tender and its subsequent evaluation marks the culmination of the procurement strategy. The evaluation of tenders needs to be carried out objectively in accordance with the developed procurement strategy. All aspects of the contractor's ability as a reputable business capable of carrying out the work should have been predetermined before inviting them to submit a tender. Evaluation of tenders should consist of comparing the submission against the context of the project requirements. The final assessment must be against the previously agreed weighted assessment criteria.

The submitted pricing schedule should be analysed against an agreed set of criteria, which should have been determined when the tender documents were prepared. Objective assessment needs to be made against:
- **Total tender price:** is the submitted price a realistic cost to carry out the work? While prices are dictated by the marketplace, too low a price can often infer a lack of understanding of the project.

- **Labour costs:** a predetermined level of dayworks and premium working should be used as a basis to cast forward these costs as the probable amount to be expended.
- **Mark-up rates for materials and subcontractors:** a schedule of expected levels of materials should be included. Mark-up on subcontractors is useful for variations and additional works.
- **Travel costs:** although travelling is the tenderer's responsibility, maintenance or serial contracts could contain a large element. Travel costs should clearly state lodging expenses and methods of travel.
- **Maintenance costs:** reviewing the contractors charges for carrying the maintenance element can be used to either negotiate a full maintenance contract or as a benchmark against future costs.
- **Whole life costs:** these are best assessed against a previously developed model. Careful analysis is required of discount rates, time periods and maintenance strategy.
- **Alternative options:** these require careful reviewing, both commercially and technically. Although a lower capital cost may be offered, operating costs may quickly diminish the savings.

The level of tender assessment to be carried out will be determined by the work package complexity. If the works package tender contained full technical requirements and specifications details, a simple cost comparison may be made on the various offers. Providing all tenders are of equal comparison technically, then there is little argument that the lowest price should not be accepted. If only the general specification and performance requirements were contained in the tender documents, the proposal must be given a detailed technical evaluation. Only once the technical and quality aspects of the tender have been considered and accepted, should it be further evaluated for programme and general management approach. The final assessment should be cost.

Rarely will it be found that all tenderers have submitted fully acceptable and compliant bids. Any exclusions, objections or differences from the statement requirements need to be assessed. This can prove difficult, as they will contain a mixture of technical, managerial and commercial differences. Reducing these to commercial amounts only, by assigning notional cost values to the technical and management differences can provide a quick assessment. If the differences are minor then they should be carried forward for final negotiations. If major differences appear then an individual assessment needs to be carried out to their acceptability. If they prove unacceptable then the tender should be set aside and discarded from future consideration.

....

Case Study

The Procurement Guidance No.3: Appointment of Consultants and Contractors produced by HM Treasury is aimed at procurement activities by government departments. A part of the tendering and award process suggested in this document is to establish:

- award criteria
- weightings for award criteria
- a quality/price ratio
- an award mechanism
- a bid basis
- price scoring
- quality threshold.

Award Mechanism: The Treasury document provides a structured approach to evaluating tenders. Tenders are assessed on how well they satisfy the award criteria (including mandatory components). The relative importance of each award criterion is established by giving it a percentage weighting so that all the weightings equal 100%. An example is given in Table 2 of the document using the award criteria, the weightings applied to them, the quality price/ratio and the price scoring mechanism to allow the quality and price element of each bid to be evaluated.

Quality threshold: The document also states a concept called the quality threshold. The quality threshold is the minimum score required in the quality evaluation necessary for a bid to be considered further. Where a bid is non-compliant because it falls below the required quality threshold, the price element of the bid should be disregarded.

...

Evaluation Interview

Depending upon the number of tenders and the selection methodology, it is important to hold tendering interviews. While they can be used for further submission and assessment of information, their main purpose is to assess the quality of intended personnel and allow the tenderer to elaborate on their approach to the project. The interview panel should consist of the client, professional advisor, designer (if used), primary contractor and the architect.

Key personnel that should attend include:
for contractors:

- project manager
- designer (if required)
- information coordinator
- lead foreman or supervisors
- commercial manager
- project engineer
- director responsible for the project
- representative's key subcontractors.

for designers or specialist engineers:

- lead designer
- project director
- designers
- information coordinator
- subconsultants.

The interview should cover the following points:

- project appreciation
- programmes
- management strategy
- staffing levels and methods
- design methods
- co-ordination
- work with other parties
- clarification of any tender items.

Any deviations from the written submission and those stated in the interview should be addressed formally. If required the tenderer should submit in writing clarification of any discrepancies.

REFERENCES

1. BSRIA (1998) *Project Management Handbook for Building Services, AG 11/98*, Bracknell, BSRIA Publications.
2. HMSO (1999) *The Procurement Guidance No.3: Appointment of Consultants and Contractors*, London, HM Treasury.
3. JCT (1989) *Code of Procedure for Single Stage Selective Tendering*, London, NJCC Publications.
4. RICS (2000) *Building Services Procurement - Guidance Note,* London, RICS Publications.
5. Wild, J. (1997) *Site Management of Building Services Contractors*, London, E&FN Spon.

FURTHER READING

The government Procurement Guidance notes can be downloaded freely from the Office of Government Commerce website: www.ogc.gov.uk

Chapter Sixteen

The Procurement Process

16.1 THE DECISION MODALITY

This final section brings together all the issues raised within the previous chapters and culminates in a single process for determining the best manner in which to procure building services. It incorporates references at appropriate places to the various other modalities so far presented.

The decision process is linear in that each section should build upon the previous. However, given the uniqueness of each project, this can never be a fixed rule. Certain projects, due to a critical project objective, may demand that the process is based upon modularisation and therefore this section would take precedence in both importance and timing over the others.

The modality is written from the perspective of the end client and is based upon a series of questions, with appropriate answers being supplied. Unlike other decision matrixes it does not end with a definitive answer. Instead, the general answers given to each question should be used to determine the most appropriate strategy for building services procurement. The wide variance in work packages, main project environment, procurement forms, contract conditions and specialisation means that strategy is possible.

The process begins with the decision over the general approach to building services, as to whether it is to be dealt with strategically by the client becoming directly involved with the supply chain, or tactically through the discharging of all responsibility to other team members. The main project procurement strategy may already exist and therefore some questions may be redundant. However, by reviewing the probable answers to the decision questions, the limitations of the chosen strategy become obvious.

Prior to undertaking this exercise, checks should be made that the building services requirements and project objectives are fully understood. Furthermore, that the client has no predetermined agreements with companies or standing polices that may affect a free decision.

16.1.1 The Initial Project Brief

For most projects briefing will be carried out in a two-stage process. The first stage sets out the headline items of the brief, and determines the requirements in the general parameters of time, cost and quality. Once the principal members of the building services team have been chosen, their expertise can be used to provide the additional detailed functional requirements to develop the headlines into a consolidated design brief.

Consider the project in question and determine the following:

General Description

- site location and features
- environment /weather factors

Project Objectives

- user needs
- business objectives

Project Brief

- purpose/function of the project
- accommodation schedule
- quality standards
- operational requirements
- specialist services/equipment
- occupancy levels and patterns
- maintenance requirements
- business criticality demands
- environmental factors
- disposal criteria
- statutory requirements
- building/services lifespan

Constraints

project

- planning conditions
- listed building status
- utilities availability

corporate

- company policies / standards
- design standards
- undertakings given

Controls

- financial budget
- timescales and milestones
- acceptable risk level
- quality standards
- cash flow

Prioritisation

- cost vs. time
- quality vs. cost
- time vs. quality

Occupation

- facilities management
- maintenance
- handover
- commissioning
- operation
- timescale

Figure 16.1 Project Briefing Headlines

16.1.2 The Project Strategy

The initial step is to determine the importance of building services on the project as a whole, and as to whether the precise requirements have been sufficiently developed to allow for objective decision making.

Key Consideration	**Probable Answer**
1. Does the building services element account for more than 30% of the project cost?	yes / no
2. Is the proper functioning of the building subject to rigid environmental conditions? e.g. temperature, air quality etc.	yes / no
3. Are whole life costs more important than initial capital cost?	yes / no
4. Do you have a predetermined preference as to who designs, installs the system or supplies major items of plant?	yes / no
5. Do you have established or probable buying power within the building services industry?	yes / no
6. Does the building services programme determine the overall project duration?	yes / no

If answering yes to more half of the above questions then building services should be dealt with as the principal issue in the main project procurement strategy.

More importantly, have the following items been precisely identified:

7. Have the requirements been precisely identified in terms of function?	yes / no
8. Have the cost limits been identified for capital and operating costs?	yes / no
9. Have the quality levels been set, together with an objective manner in which they are to be assessed?	yes / no
10. Have initial maintenance strategies been developed?	yes / no
11. Have time limits been predetermined for both design and construction?	yes / no

If answering no to at least 2 of the previous 5 questions, then sufficient understanding of the requirements has not been developed.

Decide whether sufficient expertise exists within the current project team or consider the use of a building services advisor to advise on and develop an objective set of project requirements. Does the current project team have sufficient knowledge to advise on:

12. Inputting capital and whole life cost requirements into the business case?	yes / no
13. Options appraisal of system requirements?	yes / no
14. Current industry methods and operations that will affect project delivery and procurement?	yes / no
15. Detailed assessment of project risks?	yes / no
16. Requirements to assess competency under the CDM Regulations?	yes / no

17. Able to put forward a facilities management strategy? yes / no
18. Give objective and unbiased advice on procurement strategy? yes / no
19. Give objective and unbiased advice to the best contract option? yes / no

16.1.3 Project Environment

The project environment is determined by the main project strategy. The manner in which the project principals of main contractor, architect, engineers and principal specialist contractors are procured will have considerable influence on building services procurement.

Although the main project procurement may already be determined, it is worthwhile to review how this choice will affect the decision over building services. The decision is mainly based on the three principal criteria of cost, time and quality.

Key Consideration	**Probable Answer**
Which of the following set of statements best match your project objectives?	
Option 1	
1. Do you wish to have separate design and construction organisations?	yes / no
2. Is cost accountability important or necessary?	yes / no
3. Can requirements / design be determined without reference to supply chain?	yes / no
4. Are your requirements likely to change over the course of the project?	yes / no
5. Have the design and construction been programmed? And are the dates acceptable to you?	yes / no
6. Is capital cost the only cost criteria you need to make judgements on?	yes / no
7. Are your building services requirements simple or of a standard nature?	yes / no

If you have answered yes to questions 7, 9, 11, 12 and 13 then a traditional procurement arrangement is suitable.

If you have answered no to all questions then consider Construction Management or Management Contracting or a Nominated Subcontractor under a traditional arrangement.

Key Consideration	**Probable Answer**
Which of the following set of statements best match your project objectives?	

Option 2

1.	Do you wish for design and construction to be undertaken with a single organisation?	yes / no
2.	Is cost accountability important or necessary?	yes /no
3.	Can requirements be determined without reference to detailed planning or design exercises?	yes / no
4.	Are your requirements of a prescriptive nature and easily assessed for compliance?	yes / no
5.	Is it acceptable to start construction prior to full design details being established?	yes / no
6.	Is capital cost the only cost criteria you need to make judgements on?	yes / no
7.	Do you wish to transfer as much risk as possible to other parties?	yes / no
8.	Do you wish to transfer as much responsibility as possible to other parties?	yes / no

If answering yes to the above questions then consider using either Design and Construct or a Nominated/Named Design and Install Subcontractor under a traditional arrangement.

16.1.4 The Business Case

Business Case Assessment

The business case will underpin most decisions regarding the scope of work, balance of cost between capital and operating cost, the maintenance strategy and allocation to risk. The initial decisions made within the four modalities will influence all procurement decisions.

	Key Consideration	**Output**
1.	Business Case - use modalities in Section 4.2 to determine the cost balance between capital and whole life cost	cost balance
2.	Briefing - use modalities in Section 4.3 to determine the design parameters and systems required	framework for brief
3.	Contract - based on the overall project strategy make an initial assessment of possible contracts using modalities in Section 4.4	initial contract selection
4.	Risk - identify the major risks, sources of risk and extent to which it is desirable to transfer risk using the modalities in Section 4.5	risk assessment

16.1.5 The Project Organisation

The project organisation is largely set by the approach to design and installation. With building services a number of strategic options exist, as most contracting organisations undertake some form of design service. What must be decided is what work packages exist and the overall ability to manage. Each must be reviewed as to the best option for design and installation.

Having decided that building services must be dealt with strategically, then the exact nature in which they are to be procured needs to be decided upon.

As building services can be procured as a single service or as a multitude of work packages, each must be considered as to the most appropriate strategy. Typically, decisions must be made over the principal packages of:

Mechanical

- ducting
- insulation
- controls
- commissioning
- O&M manuals/record drawings

Electrical

- lightening protection
- fire alarms
- security

Public Health

- chlorination

Specialist Services

- BMS/controls
- substation
- high voltage switch gear
- security
- data
- telecoms
- fire protection / alarms
- generators / UPS
- kitchens and cold rooms
- process/medical gases
- commissioning.

Work Breakdown

Consider the following statements and decide the extent to which each work element is to be broken down to form the principal members of the building services team:

Key Consideration		**Probable Answer**
1. Design:	Fully designed by independent party	yes / no
	Partially designed between two or more parties	yes / no
	Designed and installed by same party (or parties)	yes / no
2. Installation:	All by one company	yes / no
	Separate electrical and mechanical	yes / no
	Separate mechanical, electrical and specialists	yes / no

3.	Specialists:	Does the project have elements that can only be designed/installed by specialists	yes / no

If answering yes to 3. Then consider each work package requiring a specialist and determine the method of engagement.

4.	Scope of work	For design only	yes / no
		Two stages for design then installation	yes / no
		Install only	yes / no
5.	Method of engagement	As a full member of the building services team	yes / no
		As a second-tier supply via another principal	yes / no
		As a direct contract separate from the building services team (commissioning?)	yes / no

Detailed Considerations

Designers

1. By which method do you wish to appoint the designer?
   a) a performance specification written by the design engineer in which the contractor develops a full design to meet the stated requirements; — yes / no
   b) a partial design, with the contractor engaged to complete the detailed design; — yes / no
   c) a full design, with the contractor engaged only to install, with minor design responsibility for elements such as valves. — yes / no

Contractors and Major Suppliers

2. Consider each major work package and supply item and decide their level of influence on the project based on the criteria of cost and risk.
   a) commodity purchase - low cost, low risk
   b) cost-based selection - high cost, low risk
   c) risk-based selection - low cost, high risk
   d) strategic partners - high cost, high risk.

Competency Assessment

Follow the procedure set out in Figure 11.4 and review each possible team member for the following:

1. Do they have a sound track record of completing project to time, cost and quality? — yes / no
2. Do they have previous experience with this type of project and technology? — yes / no
3. Do they have adequate resources to undertake the project? — yes / no
4. Do they have sufficient senior management to manage the project? — yes / no
5. Do they have sufficient resources capacity to undertake the work? — yes / no
6. Do they have sufficient levels of tradespeople (or technicians for designers) and support staff that are suitably qualified? — yes / no

7. Do they have a quality policy?	yes / no
8. Do they have a dynamic relationship with their supply chain?	yes / no
9. Do they appreciate and understand risk management?	yes / no
10. Does the financial assessment show that they are solvent and financially sound?	yes / no

16.1.6 Whole Life Assessment

Whole Life Assessment

The whole life cost of the system will exceed the capital cost by a likely factor of 10. To decide the priority during design consideration must be given to the whole life.

Key Consideration	Probable Answer
1. Will the building incorporate any elements that are likely to be at risk from technological obsolescence in the next 15 years?	yes / no
2. Do any of the intended systems rely on specific technologies, chemical processes, fuels or manufacturer's specific items.	yes / no
3. Does the building have a minimum and maximum lifespan?	years
4. Is the building use dependent on a specific business process, social condition or project?	yes / no
5. What are the probable lifecycles for the principal components:	
– air handling units	years
– boilers	years
– chillers	years
– condensers	years
– control systems	years
6. Are there set limits for both capital and operating costs?	balance and priority
7. Are there pre-established service agreements that are likely to be carried over or extended to this project?	yes / no
8. Is it desirable to transfer the risk of operating and maintaining to another organisation?	yes / no
9. If answering yes to the last question, has an assessment of likely increased cost through the use of service-base agreements been undertaken?	yes / no

16.1.7 The Management Framework

People Assessment

Ensuring that the correct people and organisations are procured is paramount to a successful strategy. Consideration needs to be given as to what criteria will be used to assess appropriate parties and how their performance will be measured.

Key Consideration	Output
1. Supply Chain Management: use modalities in Section 12.2 to determine the desirability of each organisation and as to how they are best procured.	supply strategy
2. Partnering: use modalities in Section 12.3 to review the managerial approach to the project.	management strategy
3. Key performance indicators: can be used to assess the stated performance of a company and set the performance targets for the project.	performance benchmarks

Quality Assessment

To maximise the quality level within any installation, certain minimum criteria must be fulfilled. The procurement strategy must respect and include the following criteria:

Key Consideration	Probable Answer
1. Has sufficient time for the provision of a fully integrated and co-ordinated design been allowed?	yes / no
2. Has sufficient time been provided for contractors' preparation of the works?	yes / no
3. Can the works be satisfactorily executed within a realistic contract period?	yes / no
4. Do the management systems incorporate good time management principles?	yes / no
5. Is there adequate funding for design and construction and realistic tender pricing?	yes / no
6. Have adequate resources for the preparation of the client's brief, co-ordination of building services and building construction been provided for?	yes / no
7. Is there adequate management of resources evident?	yes / no
8. Have competent design and construction participants of the right calibre and experience been recruited?	yes / no
9. Is the quality of design and construction based on the client's need, with a TQM approach?	yes / no
10. Is there effective design management of the integrated design team based on TQM?	yes / no
11. Is there effective and flexible project management during the construction stage based on TQM?	yes / no
12. Have steps been undertaken to create team working with good leadership and short-or long-term partnering relationship?	yes / no
13. Is there effective and structured management of project information as covered in CIBSE's Quality Management System (AM9: 1993)?	yes / no
14. Is the contract unambiguous with fair contract terms and have clear responsibilities?	yes / no
15. Is there fair and clear risk allocation?	yes / no
16. Is there a commitment to project success?	yes / no

17. Is trust and respect evident with al; team members? yes / no
18. Has team building for improving project performance been undertaken? yes / no

Answering no to any of the above questions may lead to compromises in project quality. Answering no to more the 4 questions shows a serious deficiency in the procurement strategy.

16.1.8 Product Delivery

Product Assessment

Ensuring the end product will meet the project's objectives and client's needs is obviously paramount. Beyond the consideration of the functional needs, a number of decisions must be made as to how best the product can be delivered, i.e. through the use of standardisation or prefabrication, or can tools such as valuc engineering enhance the product.

Key Consideration	**Output**
1. Prefabrication: use modalities in Section 8.2 to determine the desirability of using prefabrication and modularisation.	supply strategy
2. Whole Life Costing: use modalities in Section 8.3 to determine the likely wlc of the design.	wlc study
3. Total Quality Management based on the procurement strategy undertake a review of the benefits of using TQM or a similar managerial process.	management strategy
4. Value Engineering: identify the major steps and probable benefits of undertaking a value engineering study using the flow chart in Figure 8.6.	risk assessment

Having considered the principal areas, the procurement strategy for the building services element of the project can now be developed.

Index

9781138372498.